Worlds Apart

By the same authors:

H. S. D. Cole et al — *Thinking About the Future*

S. Cole — *Global Models and the International Economic Order*

S. Cole & Lucas, H. — *Models, Planning and Basic Needs*

J. Clarke & Cole, S. — *Global Simulation Models*

L. Acero, S. Cole & H. Rush — *Methods for Development Planning*

C. Freeman, M. Jahoda with I. Miles, S. Cole & K. Pavitt — *World Futures: the Great Debate*

I. Miles — *The Poverty of Prediction*

J. Irvine, I. Miles & J. Evans — *Demystifying Social Statistics*

I. Miles and J. Irvine — *The Poverty of Progress*

J. Gershuny and I. Miles — *The New Service Economy*

J. Bessant and S. Cole — *Stacking the Chips: A Microprocessor Revolution and the World Economy*

Worlds Apart

Technology and North-South Relations in the Global Economy

Sam Cole

Professor, Department of Environmental Design and Planning, State University of New York at Buffalo

Ian Miles

Senior Fellow, Science Policy Research Unit, University of Sussex

ROWMAN & ALLANHELD

First published in Great Britain in 1984 by
WHEATSHEAF BOOKS LTD
A MEMBER OF THE HARVESTER PRESS GROUP
Publisher: John Spiers
Director of Publications: Edward Elgar
16 Ship Street, Brighton, Sussex

and in the USA by
ROWMAN & ALLANHELD
81 Adams Drive, Totowa, New Jersey 07512

British Library Cataloguing in Publication Data

Cole, Sam
  Worlds apart.
  1. Economic development.   2. Income
  distribution   3. Wealth
  I. Title   II. Miles, Ian
  339.5   HD82

ISBN 0–7108–0745–7

Library of Congress Cataloging in Publication Data
Cole, Sam.
  Worlds Apart.
  Bibliography: p.
  1. Economic development.   2. Income distribution.
  3. Economic forecasting.   4. International economic
  relations.   I. Miles, Ian.   II. Title.
  HD75.064   1984   337.′09′048   84-125439
  ISBN 0-8476-7374-X

Printed and bound in Great Britain by
Biddles Ltd, Guildford and King's Lynn

This book is for
Ben, Lera, Tobey and Yanina

Contents

Foreword by Philippe de Seynes*

Against the backdrop of the vivid debate of the seventies, the sunny discourse which nowadays greets every sign of a long awaited 'recovery' – for the time being a consumer-led upturn that may prove short-lived and inflation prone – seems naive, and certainly irrelevant to and oblivious of the sorry state of affairs which afflicted the North-South relationship before the onset of the crisis.

The dangers inherent in the indebtedness of a dozen countries tend to obscure, if not obliterate, the more enduring features of the development and cooperation problematique, the search for poverty-oriented strategies, for a reduction in inequalities, both within and among nations, in short for less 'unequal futures'.

A return to the pre-crisis status quo is not what the book of Sam Cole and Ian Miles is about. Rather it has its roots in the long 'winter of discontent' that caused the better part of the academic and other research centres to reconsider – and amend or reject – a number of the premises around which a certain philosophy had taken shape in the postwar era. The book examines the flaws of our social organization, which today more frequently erupt in violent political upheavals, within an international context severely constraining national options, political as well as economic and social. As the diagnosis unfolds, it reveals a world of paradoxes, unintended effects of decisions made or advocated, unanticipated turning points in observable trends, disputes over interpretations once seen as incontrovertible, perceived dangers in the attempt to generalize any findings.

Intellectual perplexities are compounded by severe ethical challenges which are the result of ambivalences in large sectors of the economy. For instance, in a world where growth transmission from North to

* Philippe de Seynes, former Under Secretary General of the United Nations for Economic and Social Affairs, was Director Project on the Future in UNITAR until 31 December 1983.

South continues to play a major role (though no longer recognized as the most effective engine of development) even undesirable behaviour on the part of the industrial North may have positive effects on the South, as long as a very broad transformation of their societies is not plausible. This is the case, for instance, with the extravagent consumption habits of Western countries, and even more distressingly with the frightening arms race. It is unfortunately true that, as the glaring budget deficit in the US emerges as the most dynamizing element of the world economic recovery, the role of the arms build-up cannot remain unnoticed.

In such a quandary it would be idle to pretend that methodological issues have been solved and do not, more than ever, require priority attention. Sophisticated anlysis is needed particularly by those whose research, as is the case in this book, is directed to more rapid social improvement, which postulates that the causes of poverty be understood and exposed. The book is the product of a far-ranging investigation, undertaken under the auspices of UNITAR, an UN organization not subject to governmental or intergovernmental control, and therefore enabled to carry out independent research, including inquiries into the consistency of UN grand designs and other important legislative documents. These latter are usually negotiated comprises, sometimes (as with the New International Economic Order) embodying powerful institutions not analytically elucidated, at other times (as with the International Development Strategy) based on professional studies of high quality, but disinclined to depart from conventional analysis.

Most of the time, the setting of such research must be that of a mixed economy with an important market sector. This is the regime under which the vast majority of economies operate, and if any attempt to improvement is to succeed, in the direction of income redistribution, satisfaction of basic needs, general welfare of the masses, and the loosening of dependency from the North, the mechanisms associated with the markets must be closely identified. Otherwise, the nature and areas of corrective intervention will not be accurately defined. This is not proclaiming a belief in the 'magic of the market', but rather in the necessity to recognize the 'power of the market', lest it defeat the best intentions of the 'voluntarists'.

Market mechanisms are not the only ones to be understood, but they lend themselves to some measure of quantification and therefore of precision and rigour in the anlysis. And, if the limitations of their explanatory power are acknowledged, these are an important asset in the development of knowledge conducive to social changes. This cognitive process derives from a blend of theoretical reflextion and empirical probing such as is aptly demonstrated in *Worlds Apart*. In the last quarter century, the blend has not been constantly of that quality, neither in respect of development policies nor the rules of international cooperation. It may have been because our perceptions of reality were somewhat dulled by the 25 years of rapid and relatively easy growth. In particular, the contribution of theory to the evolution of 'praxis', so important in the age of the classics, was often overlooked, as pragmatism, including international pragmatism, was triumphantly asserted. In the wake of Keynesianism, theoretically inclined economists were more attracted toward demand-oriented research in the context of the industrial nations. A certain specificity of Third World conditions was, it is true, recognized from the beginning. It found its formulation in the 'Two Gap theory', offering a simple, and for that reason, generally acceptable, conceptual framework to policy makers (except the most diehard devotees of the liberal school). Not the least of the product of this era was the construction of an international policy, idealistic and optimistic, which did not threaten the general configuration of the existing international order and its power structure. Much action and many new institutions resulted from that early consensus. Much progress was achieved within this framework, particularly in the emergence of a certain capacity to develop, that colonial antecedents had not generated. This was not however without a fundamental ambivalence, which was only uncovered and exposed later, when the North-South growth transmission mechanism started to break down, in the early seventies. Then, the disequalizing effects of the model, compounded by the severe demographic pressures, forced a reappraisal of the received ideas. The weaknesses of the Two Gap theory were found in its failure to analyse the full effects of international movements of goods and money, whether through trade or capital transfers, in situations of unequal exchange. The new analysis, in the development of which Professor Chichilnisky played such a pioneering role – acknowledged by the authors of *Worlds Apart* – clearly showed that these movements could not be presumed to be beneficial to the South in all circumstances, that they could have serious negative effects, possibly aggravated by the remedial measures routinely devised for their correction.

The influence of theoretical premises on general attitudes towards social problems, and consequently on specific actions of decision makers, is not easy to ascertain. They do not have to be evoked on every circumstance to work their way into the field of action. In the

national context, particularly, governments find themselves so harass-
ed by day to day problems – notably those of their balance of
payments – that it is not unnatural that they should hesitate to
question the only measures which seem to be available in the short
term, especially when they are sanctioned by a respectable tradition
and powerful institutions. Nevertheless, demonstrations that measures
recommended or imposed under crisis management might have
adverse influence on terms of trade or income distribution may at least
set the mind on new tracks, and lead to the understanding that
alternatives can exist and may be worked out over a reasonable period
of time.

In respect of the international system, on the other hand, the
influence of theoretical premises is quite visible. They are explicitly
present in the charter of major institutions, and frequently invoked in
the discussions of trade and finance, by those dicision makers and
politicians whose ideology finds support in them.

The interaction between action, theory, and ideology – the latter
sometimes elevated to the status of dogma – are therefore subtle, but
should not be ignored, in particular by those whose research is
prompted by a strong aspiration toward social change. Because they
are so keenly aware of these elements, inherent in the problematique
of modeling, the authors have successfully combined several methods
and concepts enabling them to integrate quantifiable factors within a
broader frame work bringing forth a great variety of elements,
political, social and technological, in a synthesis which offers a
penetrating and plausible representation of the world society. They
may thus usefully explore the interplay of social forces with market
mechanisms, without departing from an essential simplicity and
clarity they incorporate the most important dimensions of the debate
and development strategies, while resisting the temptations of the
percipience. Their efforts differ from other endeavours, superficially
more ambitious, by deliberately seeking to offer insights emerging
from the identification of the real determinants of poverty and
unequality, rather than relying on a wealth of statistical details, from
which historical behaviour would be extrapolated in the future. This is
obviously the right choice in the difficult problems of distribution. and
the disconcerting models of its relationship with growth.

Maldistribution is more directly and immediately felt within a
country than in a global setting. In fact, it is debatable whether the
international community should, at this stage, set for itself the specific
task of reducing inequalities in the hierarchy of countries rather than,
for instance, seeking to ensure for all the satisfaction of their basic

needs. The effect of international transactions on *domestic distribu-tion* in the North-South context is readily open to economic analyses, as exemplified in the present book. But inequality between nations has an essential *geopolitical dimension* that is not easy to ascertain, except in terms of the moral blemish suggested by its most ugly manifestations.

Yet geopolitics are present and make themselves felt in a growing number of circumstances. The debt problem is a case in point. Like the OPEC strategy of the early seventies, it confronts us with a sharp and very concrete illustration of interdependence – as an experiment in power relationship, rather than just a watchword in the rhetoric of the North-South dialogue. It comes as no surprise that the suitability of an 'OPEC of the debtors' is frequently evoked, although in this case it may well be that the position of debtors is stronger without such an arrangement, since organised solidarity might deprive them of a major weapon, namely the ever present risk to the world financial system that any one of them might *singlehandedly* opt for default. In fact a new awareness of some important facets of power relationships seems to be pervading the finance community, at least to the extent that the inclusion of the private banking sector in a cooperative international effort introduces some measure of realism and a welcome sense of a shared responsibility among the various parties involved.

Geopolitics are also very present in less extraordinary situations. They may be situations of widespread and extreme deprivation when the intellectual class takes on the challenge of poverty. Then, in the wake of violent upheavals, it becomes a matter of political necessity, and not just of ideological preference, to seek, by priority, an immediate improvement in social conditions. There is at present no adequate international response to such situations. They would require a sustained provision of liquidity to accommodate the rise in income levels and welfare programmes, until the production appara-tus has had time to adjust to the new demand conditions. Instead, they sometimes are met with embargoes and boycotts, resulting in new social commotions, with more direct foreign intervention, and the possible slide into totalitarian regimes.

Similar dangers may materialize in the very different context of the New Industrial Countries. It can now be seen that the very dynamics of their model of development, so often praised, encouraged and envied, have produced in a number of them critical disequilibria, so must so that the momentum cannot be suddenly broken. The risk of a severe – perhaps catastrophic – devalorization of assets, leading to a durable regression, may not be avoided if the econimies are not allowed to maintain a sufficient utilization of their productive capacity and the expansion of the activities it is designed to promote.

What should further be recognized – in both types of situation – and does not yet appear on the official horizon, is the need to initiate simultaneously the stop gap measures mostly designed to protect the financial system, *and* the structural actions which would correct the shortcomings of excessively outward-looking economic structures. An overemphasis on export-led industry would only prepare the early recurrence of the situation it is intended to cure. The exclusive preoccupation with financial criteria in the course of managing the crisis should yield to a concerted effort toward a new configuration of the economy, with emphasis on the output composition and a more stable balance between export sector and the expansion of the domestic market. To that effect the time-honoured dichotomy between long term and short term should be considerably loosened. The notion of adjustment, and the criteria of conditionality should be the subject of a reexamination hopefully leading to a new formulation. This process had started in a rather promising way at the end of the sixties, and was abruptly interrupted with the advent of floating exchange rates. The controlling paradigm having been somewhat shattered, it is important that it should now be reconstructed in a way which is fully suited to accommodate the acute and evolving problems of the Third World.

This debate on growth, equilibrium and distribution emerged and flourished in the seventies. It may now be seen as one major axis of reflection in the welcome renewal of the economic discipline. *Worlds Apart* is an important contribution to this movement. It offers an approach and a method which is innovative, at the same time rigorous and flexible, and provides a guide to the diagnosis of a great variety of situations. It is easy to see that its approach can be almost indefinitely adapted and refined, as well as enriched, by empirical testing, and exposure to policy makers. It lends itself to fruitful dissemination of knowledge through computer demonstrations. Although not conceived or written as a 'text book', it will soon make its way into the general learning process.

Acknowledgements

This book is an outcome of a study of *Technology, Distribution and North-South Relations*, sponsored by the United Nations Institute for Training and Research (UNITAR) as part of the UNITAR Project on the Future. The book draws on the work project team members, whose individual contributions have been published in *Methods for Development Planning: Models, Scenarios and Microstudies* (Acero et al, 1981).

Responsibility for the book rests with the present authors who have shared the drafting the various chapters. Chapter 2-4, describing the worldviews and scenarios, were drafted by Ian Miles, extending earlier work described in *World Futures: The Great Debate*. (Freeman and Jahoda, 1978). Chapters 5-8, describing out computer model and its application to scenarios, were drafted by Sam Cole. Here we acknowledge the contributions of John Clark, Ellen Evans and Tony Meagher, and especially of Graciela Chichilnisky, co-director with Cole on the UNITAR study, whose original specification of ``A Model of Income Distribution`` provided the starting-point for our model. Empirical studies were also contributed by Liliana Acero, John Bessant, Anil Date, Jay Gershuny, Raphael Kaplinsky, Bartek Kaminsky, Alfredo Nunez-Barigga, Howard Rush and Marek Okolski. The manuscript preparation, and initial editing and artwork was carried out by Vikki Razak.

The research reported in this volume was mainly carried out at the Science Policy Research Unit, University of Sussex. Office space and facilities were provided for Sam Cole while he was completing this book by the Institute for Development Studies at the University of Sussex. The hospitality and stimulating atmosphere of both institutions provided the essential environment for our completion of this work.

Finally, we are most grateful to Philippe de Seynes of UNITAR for his unstinting support for the study, and to Jean Rippert and Julien

Gomez of FUNDPAP, and the United Nations ACC working group for comments and support on the work. We share their desire for a better and a more humane future.

Introduction

CHAPTER 1

Most research into global futures concentrates on the international dimensions of distribution. We introduce a broader perspective, taking account of inequalities within countries, and recognising that these underlie conflicting responses to fundamental questions. Would the economic growth that poorest countries vitally need to redress global inequality increase inequality within their economies? Is a change in the global distribution of income possible, without changes in the power relations within each country? Is inequality in the Third World itself a result of the international system? The answers to such questions should be at the centre of global futures research today.

In thinking about the future, we argue for the use of two tools of futures studies: *scenarios* and a global *computer model*. In this study, we shall be applying them to the problematique of international and domestic inequalities. In this sense, our book is also methodological and addresses the question posed increasingly by forecasters, of how to combine the qualitative and quantiative aspects of the art.

CHAPTERS 2–4

These chapters describe the main objectives and elements of our scenario analysis. They provide a comparative analysis of major approaches to international development, including especially those supportive of a new international economic order. They set out the principal intellectual structures which are used to link national and international political change, and the economic policies and strategies. For the scenarios (and the global economic model) we view the world as a hierarchy of different types of nations, related in very different ways to the world economy. They differ in economic and

1

political strength, and in their economic structures. In these chapters, we develop our own appraisal of the theories and values involved in the various analyses and strategies that are propounded among those involved in the development debate — social scientists, policy-makers, politicians and planners, development activists and international bureaucrats.

CHAPTER 5–7

These chapters describe the model and the data used in this study. The model differs from other global computer models in several ways. Countries are not taken as homogeneous units: several income groups are represented and their interactions via domestic and international markets are displayed. The model is used to explore the redistributive impact of policies involving development aid and expanded international trade, and technological change such as the introduction of 'new' microprocessor-related techniques, or labour-intensive 'appropriate' technology, which have been key components of the various proposals for international development.

CHAPTER 8

The final chapter brings together the building blocks of the analysis in a novel synthesis of scenario construction and computer-modelling, to describe a possible 'future history' for the world political economy up to the year 2020. The chapter hypothesises a succession of global development alternatives, beginning with a continuation of the existing *status quo*, and moving successively through different development strategies in rich and poor countries. For each decade of this 'future history', we examine the prospects faced by each group of economic actors and the relationships of conflict or cooperation among groups. The chapter shows how the adoption of one strategy may follow from the limitations of another. Finally, we discuss the conditions under which more widely beneficial strategies may be devised.

1 Development, Distribution and the Future

INEQUALITY: NATIONAL AND INTERNATIONAL

Distributional issues are central to any evaluation of long-term development prospects. The way in which world output is shared between countries, and income groups within countries, is central to the assessment of development strategies in terms of their implications for the satisfaction of human needs. Economically, the level and structure of demand are conditioned by the distribution of financial resources; and, more politically, the power to influence the course of development is itself tied to the resources that can be marshalled in support of one's objectives. Unequal shares effectively mean unequal influence over the future direction of world affairs.

In recent years, a large theoretical literature and body of empirical evidence has developed around the issue of income distribution. Futures research, however, despite concern with more desirable ways of life, has tended to ignore important aspects of this work. While the future distribution of income across countries has been depicted in all but the earliest world scenarios and modelling studies, the distribution of income within countries has received far less attention, and the distribution of wealth has hardly been tackled at all. Since such aspects of distribution are key features of the world economy, in considering more desirable futures it is necessary to take them into account, using methods appropriate to handling different facets of distribution. Our aim here is to explore alternative competing strategies for global development, their implications for the national and international distribution of income over the first decades of the twenty-first century, and the trade-offs demanded of different actors in the world system.

The distribution of income within nations, and that between nations, are intimately linked through national and international economic and political exchange. One facet of these links is particu-

larly apparent at the present time of world economic crisis, when
many governments have subordinated objectives of poverty reduction
and social welfare to attempts to maintain or strengthen their position
in the international hierarchy of nations. Strategies designed to
achieve international economic and political power often do little to
relieve domestic hardship in the short or medium term; and, of course,
strategies explicitly intended to reduce poverty may be negated by the
reaction of the world polticial economy.

Higher rates of growth take on a different significance if they are
accompanied by increasing inequalities and impoverishment of the
poorest social groups. Growth and distribution have been related to
each other in quite different ways in different countries' histories. The
process which redistribute income usually involve shifts in the
balance of political power. These may reflect political changes within
countries (such as the formation of new alliances), or in the
internatonal 'rules of the game' (such as restrictions on trade or
foreign investment). They may also involve economic changes, like
those brought about by improvements in the means of production (as in
the introduction of new technologies). Competing proposals for international cooperation in development such as those which proliferated
in the 1970s place different emphasis on political and technological change. Often their proponents do not agree which political
and economic processes are responsible for desirable changes, even
when they have a good measure of agreement about what trends
would be desirable, and present their proposals as universal prescriptions. Sometimes they ignore the fact that the appropriateness of
policies and strategies depend on the circumstances of the particular
actors involved; and often they obscure the lack of consensus around
theories which purport to guide development choices. Our approach
to the future of world development will reflect this diversity of
viewpoints.

To address the issue of long-term global and national income
distributions obviously requires the use of some simplifying devices.
In this book, two analytical tools are used to unravel the web of
economic and political variables. The first is a method of *scenario
building* which takes contrasting theories of development and their
corresponding development strategies as starting-points from which
to isolate the key elements of proposals. This approach highlights the
political objectives, and the actions and reactions, of major actors.
The second tool is a *global computer model* which provides us with a
means of describing the relevant structure of the world economy.
With this, we can follow through both the economic repercussions of

individual actions, and the combined effects of the behaviour of separate actors.

In the remainder of this chapter we clarify the approach to be used and provide the background to some of the events that have led to the current state of the international economy, with its the global distribution of income.

THE CRISIS OF THE CONTEMPORARY WORLD ECONOMY

With hindsight, we can distinguish a number of epochs in the development of the world economy through the post-war years. Scammel (1980), for example, identifies three 'recognisably distinct' periods. The first, from 1945 to 1955, he describes as 'years of recovery' from the second world war; from 1955 to 1964 was a period of 'growth and comparative stability'; and the following years are described as a time of 'uncertainty and crisis'. The first period saw the implementation of the Marshall Aid Plan to revive war-torn Europe and the foundations of the Bretton Woods agreement for longer-term objectives for the international economy, together with the creation of the United Nations system — including the World Bank and the International Monetary Fund. In the second period, the recovery was consolidated, the decolonisation of many Third World nations was achieved, and economic communities were established in Europe, Latin America, the Caribbean and East Africa. For Scammel, the last period was heralded by the first United Nations Conference on Trade and Development (UNCTAD) and the realisation by the developing countries of the South that they should expect few concessions from the more industrialised Northern economies. It has seen the onset of failure of the Bretton Woods agreements in 1971 (with the revaluation of the US dollar), the shock to the already unstable international system brought about by the first oil crisis in 1973 and, the subsequent high rates of inflation in the industrial countries, severe balance of payments problems in many industrial and developing economies, and increasing rates of unemployment in most world regions.

Views on international inequality changed considerably through these different phases of world development. If there was a predominant view of the future of global income distribution in the decades immediately following the second world war, it was that international inequalities would diminish. In the post-war boom, economic growth was expected to close the gap between rich and poor countries, and lead to a reduction of the large inequalities existing in the less developed countries. The role of development policies was encapsulated in the term 'modernisation'. The poor countries were simply in an undeveloped state, and as their traditional practices were eroded by the importation of modern technology and Western tastes and values, it was foreseen that they would necessarily follow the development path established by the industrial world.

The economic policies adopted both internationally and domestically were much more complicated than this classic 'trickle-down' view suggests. The market was not left to govern post-war reconstruction by itself. Marshall Aid was, in effect, a Keynesian policy at the international level between the Western nations – although, it has also been viewed as less a matter of pure economics than as a geopolitical strategy aimed at containing the rising power of the Soviet Union. Further, even among the market economies, many European countries had strongly welfare-oriented and interventionist policies. Whether market-oriented or not, most newly-independent countries were encouraged to adopt planning as a condition for receipt of foreign aid. Consequently, while the philosophical basis for post-war policy at the international level was dominated by the position of the United States, experiments with contrasting theoretical underpinnings were in progress at the domestic level.

The stress on markets, at least as a mechanism of international distribution, was brought into question by subsequent events. By the 1970s, books were appearing with titles like *The Widening Gap* (Ward, 1970), and *The Challenge of World Poverty* (Myrdal, 1970). Figure 1.1 is representative of the data that were widely cited at the time. During the "post-war boom" the gross domestic products of both rich and poor countries increased at more or less the same rate. Given their different starting-points, this meant a much larger absolute increase in production in the rich countries, and when the much higher population growth of poor countries was taken into account also, the differences were even more apparent. Although overall economic circumstances improved in all regions, the *per capita* income growth in the poorest countries was only a quarter of that in other regions, and the difference in absolute income levels was huge and increasing. Figure 1.1 indicates the world that was expected to result if these trends had persisted into the end of the century.

Primary responsibility for the low *per capita* income growth in developing countries was widely attributed to their rapid population growth. During the 1960s it began to be argued, too that increasing populations, together with rising standards of living, would lead to a more serious ecological threat. Thus, curbing population growth was made a priority as a development issue, and some voices argued that high living standards could never be supported worldwide. Such arguments became more widespread as problems in the world economy became increasingly apparent.

Figure 1.1: *The widening gap in 1970*

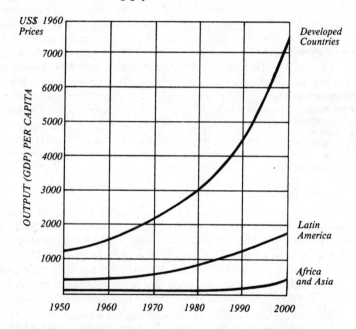

Source: Based on Ward (1970).

We prefer to follow most 'long wave' analysts, and date the onset of the present world economic crisis rather later than Scammel does; we view UNCTAD 1 as an expression of discontent with the international economy rather than as a symptom of its debilitation. The year 1970 is a convenient milestone. In the early 1970s widespread concern was voiced in the West over inflation and rising state expenditure — 'the fiscal crisis of the state'. Concern too, for the stability of the international economic order (after the US dollar was revalued all major currencies were 'floated'), and for the increasing problems of many traditional industrial sectors like steel and ship-building, became apparent at the beginning of this decade. The 1973 oil crisis accelerated the worsening trends in inflation, debt and unemployment, and marked the onset of falling growth rates in many parts of the world. (See Table 1.1).

Table 1.1: *Trends in growth of domestic product 1950–82*

Economy group(4)	Year				
	1950–60	1960–65	1965–70	1970–80	1982
Industrialised market economies	3.8(1)	5.3	4.9	3.2	–0.2
Centrally planned economies	—	6.2(2)	7.7(2)	6.4(2)	1.9(3)
Africa, South of Sahara	3.6	5.0	4.9	3.4	4.0
Middle East and North Africa	5.1	6.4	9.4	4.9	—
East Asia and Pacific	5.2	5.5	8.0	8.2	4.2
South Asia	3.8	4.3	4.9	3.2	3.5
Latin America and Caribbean	6.1	7.5	6.5	6.0	–1.2
Southern Europe	5.3	5.2	6.1	4.6	2.2

Notes: (1) 1955–60.
      (2) Weighted average of country growth rates.
      (3) Data for 1981.
      (4) Regional membership shows some variations according to source.
Source: World Bank (1980) World Tables (2nd edn); World Bank Development Report (1982), Washington, IBRD, and World Economic Survey (1981–2).

The combined power of Third World oil exporters, together with signs that other producers of vital materials were also seeking to form cartels, gave impetus to the North–South dialogue. However, many of the new initiatives soon foundered, and despite the United Nations' call in 1974 for a New International Economic Order, and other significant declarations of intent such as the Lima Declaration and the Lagos Plan, no far-reaching plan of action has been forthcoming.

From the mid-1970s, uncertainty about the world's economy

increased, and progress towards agreement on improved international relationships was harder to achieve. Shrinking world trade tended to pit countries against each other; national industries competed aggressively in diminishing world markets for their products. In an effort to shelter domestic industries from imports, and to maintain a healthy balance of payments, nations have become increasingly protectionist. Negotiations over world trade and currency matters have made little progress. To quote the United Nations Committee for Development Planning (1983, p. 3):

The threat to the international trading system is as serious as the emergency regime in the international financial system ... policies that seek to pre-empt shrinking markets are not only disruptive to the world economy but infuse bitter conflicts in international relations.

Third World countries had shared least in the benefits of the post-war boom, and UNCTAD 1 was among the most articulate early expressions of a desire to change this situation. Most of these countries, and particularly the larger oil-exporting economies, faced special problems in the worsening economic crisis. They tended to incur large balance of payments deficits as oil prices, and subsequently those of northern manufactured goods, escalated; their external debt increased; agricultural producers, dependent on fertilisers and producing few of their own manufactures, were particularly hard hit. While some (mainly Asian) newly-industrialised countries coped comparatively well — at least in terms of maintaining their industrial growth — the poorest economies fared badly. Increasing economic hardship led these less developed countries to press harder for the reform of the international institutions.

By the 1980s, even the World Bank — traditionally a staunch supporter of economic interdependence through trade — put forward a slightly jaundiced view of international exchange:

For much of the past thirty years, growing interdependence — through trade, capital, and migration — strengthened the forces of economic expansion and spread them around the world. But, as recent events have illustrated, these links can transmit problems from country to country just as surely as benefits (World Bank, 1982, p.1).

A cynic might conclude that it is only when problems begin to be transmitted *within* the North, as well as from North to South, that the inadequacy of the international economy begins to be recognised in this way.

The 'transmission of problems' in the world economy has meant that the economic crisis has had severe impacts in even the strongest industrial and developing countries. Both Japan and the Federal Republic of Germany, typically the most dynamic of the major industrial powers, began to suffer from weakened world demand for their exports and to experience the effects of recession. The same has also been true for the wealthier oil-producing countries, whose exports have fallen steadily since 1980, and also for the newly-industrialised economies (NICs) whose growth has depended on exports of manufactured and intermediate goods to the increasingly uncertain markets of the industrial world.

How are the changes in the global economic outlook reflected in analysis of long-term prospects? It is interesting to note that the upsurge of futures studies which we reviewed in *World Futures: The Great Debate* (Freeman and Jahoda, 1978) coincided fairly closely with the onset of the world economic crisis. In this literature, most of the authors are implicitly or explicitly arguing with each other about the appropriate diagnosis and prescription for the world's ills.

One line of argument in the futures literature, characterised by the work of Kahn and Weiner (1967), both prescribed and predicted the continuation of growth and stability, and carried a strong flavour of 'technological optimism'. This was followed by the more despondent *Limits to Growth* forecasts of the Club of Rome (Meadows *et al.*, 1972), foretelling both environmental problems and Malthusian resources constraints in the face of an expanding world population. Economic and demographic growth in a finite world were here identified as major problems, threatening to despoil the biosphere. Economic problems — especially the oil crisis that shortly followed the publication of *Limits* — were interpreted as early warnings of the self-defeating nature of growth-oriented policies. Later Club of Rome studies placed less emphasis on eco-catastrophes, and highlighted problems arising from the increasing complexity of the world economy (Mesarovic and Pestel, 1974).

Such concerns, paralleled in the United Nations Conference on the Environment in 1972 and Population Conference in 1974, were superseded by more direct concern with issues of global distribution, for example, by the Bariloche study of Herrera *et al.* (1976), and the United Nations World Model (Leontieff Herrera *et al.*, 1975). The studies which followed these, largely based in international organisations such as the OECD (with its interfutures Project, 1979) and the World Bank, increasingly concentrated on how to restructure the international economy so as to restore economic growth.

Figure 1.2: *Comparison of high and low projections to the year 2000*

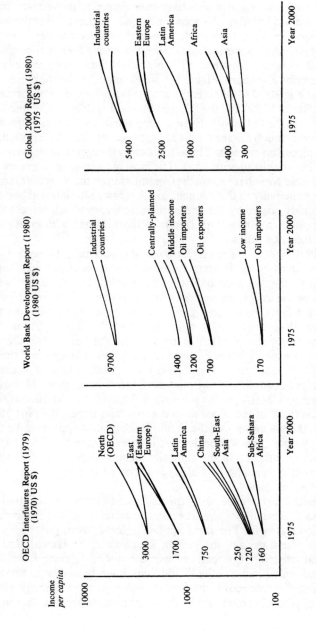

Note: In each case the highest and lowest projections or growth rates are taken. For the World Bank projections 1985–90 anticipated growth rates are used.

International organisations have been forced to move away from extrapolations based on the economic trends displayed over the post-war boom, in order to provide forecasts of development over the coming decades. Nevertheless, their quantitative forecasts of global regional economic growth often do not reflect the severity of the crisis they describe elsewhere. Figure 1.2, for example, shows the projection on the Year 2000 (1980). Even among these institutions, forecasts World Bank (1980) and the (United States) President's Commission of the Year 2000 (1980). Even among these institutions, forecasts produced at much the same time may show considerable variation, as this comparison reveals. These forecasts all suggest increasing *per capita* incomes in all economies. The most obvious long-term slowing of economic growth is seen in the Commission on the Year 2000 study — due principally to their conclusion that raw materials will become increasingly scarce by the end of the century, in which respect they have more in common with *Limits to Growth* than with most other recent futures studies.

With respect to the relative growth of economies, the projections all show gains by the Eastern European and some Asian economies relative to the Western industrial economies, but relatively weak growth by the lowest income African and Asian nations. Interfutures projects a fairly rapid closing of the average *per capita* income gap between Western (OECD countries) and the Soviet Union and Eastern European (CMEA countries). By the year 2000, the *per capita* income of Eastern Europe is expected to be at least half that of the OECD economies. The World Bank (1980) likewise forecasts more rapid economic growth for the CMEA economies. However, even though CMEA growth rates are reckoned to have declined much less than those of industrial market economies during the post-1975 'crisis', the assumption of the World Bank studies appears to be that recovery from crisis will be delayed in these economies. Even so, the long-run growth rate of the centrally planned economies (including the People's Republic of China) is expected to exceed that of the market economies.

Over the last decade, the expected growth rates for most nations have been reduced with almost every annual revision of such forecasts of regional growth. The same applies to forecasts of growth for the OECD countries, and the growth forecasts of the United Nations regional agencies. Forecasts of world economic growth from such international agencies as the World Bank, have, by and large, likewise been steadily reduced. Furthermore, although most agencies give a range of possible growth rates, there are now suggestions that in the short term only the lowest rate is likely to be achieved. In the World Bank *Development Report* (1983), projected annual growth rates for all developing countries vary between 4.7 per cent and 6.2 per cent, and for industrial countries between 2.5 and 5.0 per cent. But the World Bank's assessment is that over the next decade even a 4 to 5 per cent GDP growth for developing countries as a group would only be feasible if 'a concerted national and international effort ensures that recovery from the current recession is strong and long lasting, and that individual economies are restructured to meet the changed economic conditions.'

The economies of the industrial countries were described by the International Monetary Fund *Annual Report* (1982) as being stalled in a prolonged period of slow growth. Aggregate growth of real GNP in these countries averaged only approximately 1.25 per cent over the years 1980 and 1981, and 1982 was marked by even lower (or negative) rates of growth in major industrial countries. Forecasts of imminent economic recovery in the United States and the ailing European industrial economies have been a regular feature of governmental pronouncements since 1973 (when the economic slowdown was first widely acknowledged). But the rapid growth in the 1960s and early 1970s ended with the recession of 1974 and 1975, and partial recoveries have been short-lived. Although optimism about economic affairs is still regularly expressed by national administrations, most inter-governmental agencies expect that the crisis will only end after a good deal of global and sectoral economic and political restructuring. The United Nations Committee for Development Planning (1983 p.4) concludes:

In view of the tensions and imbalances that characterise current international economic relations, not least the towering debt problems, there can be no confidence that the recovery signs visible in early 1983 will be sustained, and still less certainty that the recovery will spread to the developing countries.

The abrogation of many of the old rules of the GATT and Bretton Woods agreements which governed the post-war period, has been reflected in a resurgence of conflict over the interpretation of economic trends and policies. Economic and political issues are now generally seen to be closely linked, and the concept of 'restructuring' domestic and international systems has become prominent. However, there has been little agreement over the shape that this restructuring should take. Divisions have sharpened within and between countries, and — as we examine in the next chapter — divergent accounts of the nature of the global problematique are proposed. The issue of international income distribution is being posed with increasing force. But the global economic crisis also throws the future of income distribution within countries into question.

INEQUALITY WITHIN COUNTRIES

Per capita income growth over the 1970s, as shown in Table 1.2, fell well below the exponential growth predicted at the beginning of that decade (Figure 1.1). The expectation that the major dimensions of the gap would remain were, however, substantially fulfilled. There are a number of reasons for caution in interpreting these statistics as measures of welfare, or even of economic power. The purchasing power of money differs considerably around the world: and when this is taken into account, the differences between world regions are reduced but still massive, as Table 1.3 shows. Furthermore, much of the output of poor countries is non-marketed, and thus not fully represented in national accounts statistics.

Table 1.2: *Trends in income per capita 1950–80*

Economy group (4)	Year			
	1950–60	*1960–65*	*1965–70*	*1970–80*
Industrialised market economies	2.5(1)	4.0	4.0	2.5
Centrally planned economies	—	4.8(2)	6.7(2)	3.9
Africa, south of Sahara	1.3	2.4	2.3	1.6
Middle East and North Africa	2.6	3.7	6.5	2.7
East Asia and Pacific	2.8	2.8	5.4	5.7
South Asia	1.8	1.9	2.4	1.1
Latin America and Caribbean	4.5	6.0	5.0	3.4
Southern Europe	2.4	2.3	3.3	2.9

Notes and Source: As Table 1.1.

Table 1.3: *Comparison of GDP estimates as obtained by conventional methods and with purchasing power taken into account*

Region	1975 Conventional GDP	Adjusted GDP
North America	1528	1520
Europe	1774	1757
Latin America (including Caribbean)	546	806
Asia (including Oceanic)	974	1471
Africa	175	324

Units are US$ billions (1975)
Source: Kravis *et al.* (1982).

Beyond this, the meaning of any particular level of income *per capita* for a country will be very different for countries with different degrees of income inequality. Countries at the same levels of national income have diverse types of inequality, which is reflected, for example, in their different performances in life-expectancy and other measures of social welfare.

Inequality, whether within or between countries, is by no means a simple matter to conceptualise or measure. The particular aspects of inequality which researchers and statisticians focus on depend upon the research questions and concerns that they bring to their work. *Absolute poverty*, for example, which might be measured in terms of the proportion of the population falling below some minimum income level, is of particular interest if we are concerned with the extent to which people's basic needs are being met. *Income distribution* is important for considering social justice, and determining how different groups are benefiting from economic development. The *distribution of wealth* has received rather less attention, but it is of considerable relevance to the ways in which inequalities are structured and reproduced, and to the location of power in societies.

There are considerable difficulties associated with the measurement of income inequality, especially when making international comparisons. Differences in measurement (e.g. personal distribution versus household distribution) and in the way in which redistributive policies are administered, make for problems in comparing statistics globally, Moroney (1978), for example, observes that alternative plausible assumptions concerning tax incidence for an industrial economy, led to changes in the after-tax Gini coefficient (one common measure of distribution) which are greater than the differences between Gini coefficients reported for the USA, UK, Pakistan and India. The data are far from satisfactory; nevertheless, researchers have sought to make the best of a difficult task.

For several decades now, there has been a strong tradition of research concerning national variations in income distribution. At least since the pioneering work of Kuznets (1955, 1963), there has been considerable interest in his hypothesis of an 'inverted U-curve'. Such a curve has been repeatedly identified in cross-national studies. The greatest degree of inequality is reported, typically, for middle-income countries (see Table 1.4).

Table 1.4: Economic development and the distribution of income

Per capita GDP 1965 US$	Number of cases	Gini Index	MEP Index	Share of total income received by: Top 5%	Top 20%	Poorest 20%
<100	9	0.42	31.6	29.1	50.5	7.0
101–200	8	0.47	37.2	24.9	56.5	5.3
201–300	11	0.50	38.5	32.0	57.7	4.8
301–500	9	0.49	37.6	30.0	57.4	4.5
1001–2000	10	0.40	29.0	20.9	58.6	4.7
>2001	3	0.36	26.2	16.4	52.7	5.0

Note: Gini and MEP are both overall inequality indicators. The Gini index would be zero if different percentile groups of the population received equivalent proportions of the national income, i.e. if there were perfect equality: it measures the area between the line of absolute equality and the Lorenz curve that plots the actual income distribution across percentile groups. The MEP is the 'maximum equalization percentage' which represents the percentage of total income that would need to be shifted between different quintiles of the population in order to achieve equal distribution over the quintiles.

Source: Paukert (1973).

This result, together with a number of studies suggesting that economic growth in some developing countries was associated with increasing inequality, meant that few commentators now argue that 'modernisation' will automatically reduce inequalities by eroding the privileges of traditional societies. The inverted U-curve, rather, has often been interpreted as meaning that the poor can expect to get poorer (at least in relative terms) before they will get richer. One common explanation for this relates to the processes of 'modernisation' — new economic sectors and occupations emerge, where (as they are generating economic growth) income levels rise more rapidly than do those in the relatively unchanging traditional sector.

More recently, this view has been contested by structuralist researchers. Some relate inequality to increased integration into the world economic system, which differentially rewards different social groups. Others argue that the relationship between modern and traditional sectors (urban and rural, in some accounts) is important, but that it is a matter of unequal 'terms of trade' which impoverish the poorest groups within countries, in much the same way as the poorest countries are disadvantaged by unequal exchange within the world economy. (A useful review and attempted integration of different approaches is provided by Bornschier (1983).)

We shall consider different analyses of development and underdevelopment in some detail in the next chapter. For the present, we make a number of observations based on the empirical evidence that are important guides to thinking about the future. The discussion above will have made it clear that economic growth in itself is not necessarily a guarantee of increased welfare for the poor. To go beyond this, it will be helpful to consider the experiences of countries at various levels of economic development.

Income distribution in developing countries

Political and social factors are important determinants of inequality: many countries deviate considerably from the inverted U-curve, which cannot be viewed as an 'iron law'. There are certainly ambiguities in the data, but for many developing countries research rather consistently suggests that growth has primarily benefited the upper or middle classes and that income distribution has become more unequal. While the absolute standard of living for some groups has increased substantially, and others have seen marginal improvements, substantial proportions of the population in some countries have experienced a decline (Ahuwalia *et al.*, 1979).

A recent review of both cross-country studies and studies of single countries over time (organised in terms of Latin America, Asian and African case studies) conclude that there is support for the inverted U-curve hypothesis as a description of present trends: 'inequality first grows and then declines as income grows', but 'the level of *per capita* income only explains a limited part of variations in the material...

other factors together are more important than the level of *per capita* income.' (Bigsten, 1983, p. 68). From time-series data, it is shown that income inequality and absolute poverty have tended to increase in Africa, while a more mixed picture is presented for Asian and Latin American countries. While data are inadequate, there is some evidence that centrally-planned countries in these regions have rather less inequality than the others, in whom middle-class and, in some cases, the top few per cent of income recipients have been the main beneficiaries of growth.

In a more detailed survey of experience in thirteen Third World countries, Fields (1980) considered both distribution and poverty. In only five of his thirteen cases (Costa Rica, Pakistan, Singapore, Sri Lanka and Taiwan) were both poverty and inequality reduced. There was no clear link between more or less rapid growth and shifts in equality. Sri Lanka was one of the slowest growing cases – but so was India, where distribution improved but poverty did not. Taiwan had relatively high growth – but so did the Philippines, where absolute poverty remained high. Government policies clearly played a role in the divergent trends.

Experience among the newly-industrialising countries is also seen to be diverse. Bigsten (1983) reports increasing inequality in Latin America NICs, but better distributional trends in their Asian equivalents. Some authors (e.g. Ranis, 1983) suggest that this can be related to export-oriented development strategies: others (e.g. Evans, 1983) relate it more to the geopolitical location of the Asian economies which led, for example, to reconstitution of wealth inequalities in the form of land tenure reform, and to particular influxes of foreign aid and of skilled immigrants.

For centrally-planned developing economies, there is typically rather little data on which to base a firm judgement, and it is difficult to speculate about what trends are taking place in these countries. Bigsten (1983) concludes that the extent of poverty is smaller, and the degree of inequality less, than in other countries of a similar *per capita* income level. Fields (1980), too, suggests the limited available evidence shows that the 'newly socialised' countries have achieved greater equality than non-socialist countries at similar stages of development. He remarks also that the growth records of the socialist countries in his sample are 'far from exemplary'.

Income distribution in industrial countries
For industrial countries, especially those of the West, more frequent and reliable data are available. This does not always make interpreta-

tion easier, since detailed comparisons reveal a number of inconsistencies! It is again clear that the experiences of individual countries are quite diverse, and that even those underlying trends that have been reported are typically subject to a variety of fluctuations. Some earlier studies (e.g. Soltow, 1968) suggested that there have been very long-term trends towards greater equality, spanning several centuries, and it is generally recognised that the twentieth century has witnessed major increases in the equality of distribution of incomes within western countries — much of this to do with redistributive policies associated with the establishment of 'welfare' states in the post-war reconstruction.

Evidence for trends in income distribution in western countries over the post-war period suggests rather varied developments: We conclude that rather little change in distribution occurred over the course of the post-war boom. Assessing the decades before the present economic crisis really began to bite, for the United States, Moroney (1978) reports no perceptible long-term trend; although Chapman (1978) suggests that inequality decreased between 1939 and 1949, and then subsequently increased from 1959 to 1968. A comparative study by the Royal Commission (1973) suggests that there was a trend towards inequality after 1967, although Chapman seems to disagree. The Royal Commission concluded that an uneven trend towards greater inequality is perceptible for the United Kingdom after 1967 (preceded as in the United States by a decline between 1959 and 1967, that estimates for Japan, too, indicate a long-run trend towards inequality. For West Germany, since 1955, they found little change, while in France inequality increased between 1956 and 1962 and then decreased until 1971.

Most of these data cover the period up to 1975. More recent data on income distribution are in short supply, but since this has been a period in which unemployment has increased dramatically in most countries, it seems likely that income distribution must also have worsened. Certainly, the available data on poverty suggest this to be the case. For the United States, the 1982 population census data show the number of people below the poverty line to be the highest since 1965: in 1981 alone, the poverty register was extended by 2.2 million (Thurow, 1983). In the United Kingdom (DHSS, 1983) data show that between 1979 and 1981 the number of people living on the margins of poverty increased from 11.5 million to 15 million, more than one quarter of the total population.

How do these countries compare in terms of inequality? Estimates by Sawyer (1976) suggest that among the industrial market economies, greatest equality has been achieved in the Netherlands, Norway, Sweden; ahead of the United Kingdom, Japan, and even further ahead of the United States, France, Germany and Spain. These differences correspond with the social welfare orientations of the countries concerned, a point which has been demonstrated by Cameron (1978) and Stephens (1979), who have related equality (and welfare state expenditure) to the strength of social-democratic and socialist parties across Western countries. It is less clear that distribution is related to growth (although Cameron suggests a negative relationship), and there is no evident association with national income among this limited set of countries.

Apart from explicitly redistributive policies, several other factors contribute to the present pattern. The Royal Commission (1973), for example, suggested that the comparatively high proportion of wages and salaries in total personal incomes accounts for the relatively favourable performance of Sweden and the UK, and the greater proportion of rent, interest and dividends for the poorer performance of the USA. They suggest also that demographic factors (in particular, the age structure of the population) and the labourforce participation rate (especially that of women) are important.

What of the more industrialised centrally-planned societies? There is again some evidence that these countries tend to be more equal than most of their western equivalents, in large part due to the absence in the Second World of the affluent classes whose income derives from profits and rents (Lane, 1971).

Moroney (1978) suggests that post-tax distribution in the UK is comparable so that in the smaller industrial socialist economies (Czechoslovakia, Hungary and Poland), and better than in the Soviet Union. Indeed, until the 1960s, earnings inequality in the Soviet Union was comparable to that in the United States. Moroney (1978) concludes that the differences in income inequality across socialist countries, and that changes in the Soviet Union, especially since the last world war, have been at least as great as the differences between the socialist countries as a group and the United States. Of course, there are differences in both national income levels and demographic characteristics between the First World and Second World countries, but it seems reasonable to conclude that only fairly minor differences in the levels of income distribution exist across these regions. However, other price and welfare policies may significantly increase the access to economic (if not political) resources of poorer groups in the ·Second World as compared to people of comparable income levels in the West.

The complex relationships outlined here cannot be explained in terms of the simple inverted U-curve hypothesis. If anything, national income level is only related to income distribution by virtue of a range of other associated variables: the sectoral distribution of the labour-force, the levels of economic activity of the population, demographic structures, and location in the international economy. We can relate these factors to the supply and demand of labour of different types, the balance of which will tend to influence the wages offered to various categories of workers.

Other variables, however, of more directly political form — fiscal, welfare and educational policies, for example — affect both distribution and access to resources. Analysis of political factors, as opposed to particular policies, is not very well advanced. Thus Bigsten (1983) reviews the evolution of Third World income distributions and then notes that:

One problem upon which I have only touched briefly . . . is the question of political power, and it may well be the crucial factor. Drastic changes in a direction that favours the poor cannot be expected, unless they obtain a certain amount of political clout. (p. 96)

The distribution of wealth
An important determinant of political power is access to economic resources of wealth. However, the empirical study of wealth is made difficult because control over resources may be obtained by means other than their formal ownership. The ownership of land is particularly important in Third World countries, and there is considerable evidence that land reform is an important step towards relieving rural poverty (Griffin, 1981).

For other economic resources there is a paucity of data for comparative analysis. Data presented in Chapter 5 suggests that income from investments is less well distributed in developing economies than industrial economies. Western countries show some signs of decreasing inequality: For the UK, Atkinson (1973) reports a slight decline in the share of wealth of the richest 5 per cent of the population (from 80 to 70 per cent between 1920 and 1950). A similar pre-war improvement in wealth distribution in the USA (associated with the stock market crash in 1929) has slowed or stopped. In so far as a comparison can be made, Atkinson suggests that wealth is less concentrated in the USA than in the UK. But as Kay (1979) observes, despite the exceptionally high estate duty in the UK, little redistribution of wealth has been effected.

However, the effective control of wealth by means other than direct owernership means that we must treat such data with caution. It may well be that with increasing concentration of economic activity in giant corporations, the power of very small portions of the population is actually continuing to grow.

Similar problems apply to centrally-planned economies, where major economic resources are generally under state ownership. It is by no means evident that their state agencies and enterprises really act as if they were 'controlled by the people'. Rather, it would seem that functionaries exert considerable power with very little public regula-tion – as in many Western state agencies. It is difficult to estimate the distribution of effective control over resources in either type of economic system.

Thus, data on wealth distribution are even less satisfactory than for income distribution. But the two aspects of economic inequality are highly related — the inheritance of wealth is an important determinant of individual incomes, and the overall shape of income distribution may be affected by the distribution of wealth. Policy interventions, such as land reform and the confiscation of property, make drastic changes in the distribution of wealth. But we suspect that, in general, command over economic resources tends to change comparatively slowly across social groups compared to income distribution.

Since both economic and political factors are involved in determin-ing these inequalities, our forecasting approaches will need to be able to take both into account. In practice, the model we shall introduce involves a number of assumptions which relate wage levels to the supply and demand of labour, but distribution can be varied according to prospects of political change. These political changes are outlined on the basis of our scenario building.

From this preliminary appraisal of distributional issues in the contemporary world, we now turn to a more methodological discus-sion. The remainer of this chapter outlines the two forecasting methods which we shall be bringing together in our analysis: scenario construction and computer modelling. Subsequent chapters will develop and apply these methods to the issues we have been discussing.

SCENARIO CONSTRUCTION

The scenario approach to forecasting was first identified formally as a distinctive methodological tradition in the 1950s, when it began to occupy the attention of military planners. In making plans for different contingencies, they found it useful to identify the ways in which different events might relate together. In a gaming exercise, for instance, the participants would be assigned the roles of personnel in different armed services, and asked to imagine that they were being faced with choices in a particular hypothetical situation. The 'scenario' would typically be the description of this situation and the events leading up to it, and their own decisions could enter into a subsequent scenario.

In many ways, this sort of speculative 'future history' was nothing new. It has long been at the heart of science fiction and social philosophy. Even the much vaunted 'systems thinking' that was proclaimed as a specific advance of post-war policy analysis has its precursors; and we might be tempted to conclude that the 'systematic' approach claimed by many recent futurologists is more a matter of scientism (appealing to scientific authority) than of particularly rigorous thought. Nevertheless, the term 'scenario' was popularised through its use in discussions about the risks of nuclear war, and in the increasing use of scenarios in corporate planning. The term achieved prominence with the late Herman Kahn and Wiener's *The Year 2000* (1967), which dazzled readers with its wide-ranging speculation about the last decades of the twentieth century and the social and technological accomplishments these could display.

Kahn's futures were based upon trend extrapolation, together with some use of Delphi forecasts of possible technological breakthroughs. His different scenarios were, in the main, relatively minor variations of detail upon trends in economic growth and technological capability. Much of the fascination of the book came from the verve with which the trend extrapolations were converted into startling discussions of tomorrow's world. As well as providing vivid examples of scenario presentation, Kahn supplied a useful definition of a scenario as a 'hypothetical sequence of events constructed for the purpose of focusing attention on causal processes and decision points'.

There are a number of virtues to this definition. First of all it implies that we are not just interested in trends, but also in the forces that underlie them. Understanding these forces can help us reach conclusions about whether the trends are likely to continue, and what factors might bring about discontinuities. Secondly, it directs attention towards decisions, to the role of human choice in creating the future, and to the possibility of assessing when and where a given action will have a particular effect. Thirdly, it implies that we should be concerned with *alternative futures* rather than with *the future*. If there are key decisions, or potential 'branching points' where chance or human agencies may play a decisive role, it is important that these can be taken into account in forecasting. The scenario approach is, indeed, often seen as being synonymous with multiple scenario analysis.

The subsequent chapters will elaborate our approach to scenario building. Our scenario analysis includes the use of the computer model to explore some of the key decisions and branching points that are implied in the scenario construction. However, a few points about our approach can be made here with reference to some previous studies in this field.

From the discussion above, it is apparent that we are not using the term 'scenario' to refer to restricted listings of economic or demographic trends. We consider that the term should apply to accounts that encompass both events and the actors of social groups whose decisions may influence the future. This approach — along with our concern with distributional issues — requires a specification of different interest groups. This is common in futures studies, although the main reason for identifying social groups has often been simply to depict the impact of change upon them. For example, nation states have been the main unit of analysis in most global studies. Authors may even discuss prospects for social groups within countries: Kahn also speculates about the ways of life of rich and poor people in different countries (and in later work discusses the situation of Third World peasants and urban workers); Interfutures (1979) (which also related a model to some of its more qualitative work) was also mainly concerned with scenarios for countries of different types, but this study did discuss different social groups when considering value changes which might reorient development in Western countries.

We take different social groups as actors, as agents capable of producing social change. Given our concern with distribution, different social classes are the main actors within countries that we shall consider. Social classes are groups described in terms of their access to wealth, their different degrees of control over the means of production of economically significant resources. While there may be many such groups within a country, some simplification will be required in our study. In the modelling work, in particular, we shall represent social classes in each economy in terms of two or three income groups. Our qualitative discussion can be somewhat more detailed, as is illustrated in Chapter 4.

One final point needs to be signalled in advance about our scenario approach. We have already noted the diversity of explanations of distributional issues, and in this study we distinguish between three worldviews — three distinctive and substantively different ways of explaining the processes of world development. *Conservatives, reformists* and *radicals* propose very different analyses of why the world is as it is, and how it might be changed for the better. Our own values and commitments will be more clear if we relate them to a framework provided by these different worldviews. Thus, readers will be able to identify how far our conclusions are dependent upon assumptions with which they may not agree. It is, in any case, fruitful to contrast the different worldviews and gain some idea of how far each can help us grasp the complex reality of world development. Significant social groups in the world today are committed to different worldviews; they believe them to be useful descriptions of 'reality', and these in part underlie their strategies and visions of the future. By differentiating between worldviews, we can achieve a better grasp of the likely patterns of development should one or other group be empowered.

We shall put the philosophy guiding our scenario building into practice in the following chapters. We now turn to the approach we adopt to computer modelling, and how this relates to the scenarios.

MODELLING AND SCENARIOS

The techniques of scenario building and computer modelling used in this book are complementary. For the purpose of examining the issue of global income distribution, it is especially important to interrelate economic and sociopolitical behaviour. The approach we develop in this book helps us to blend these ingredients.

The *Limits to Growth* study (Meadows *et al.*, 1972), sparked off the present interest in global models. In summarising the Sussex critique of *Limits,* Jahoda (1973) argued that it failed to take account of the 'human factor', and in particular, that the model had no mechanisms to describe people's possible reactions to the dramatic events which it predicted. Computer models tend to treat patterns of behaviour as fixed, or following historically defined patterns. Further, since they require, ultimately, that all information is translated into a numberical form, this usually sets limits on the kind of variables which may be included into a model. But there are reasons why we should not simply abandon formal modeling (see Clark and Cole, 1975). Thus Bigsten (1983), after reviewing the literature on distribution, argues for the value of modelling. 'Even if sociopolitical factors are very important in this field,' he notes, 'it seems to me that more methodological and theoretical stringency is required, if we are to proceed beyond the present level of understanding' (p. 59).

Most importantly, computerised and other formal models have the tremendous advantage that they can help us to follow through interactions of particular variables and assumptions, according to a particular and explicit logic. Even though the precision achieved is sometimes more apparent than real, the formalisation of assumptions forces a rigour and consistency difficult to achieve by other means. Also, of course, the computerisation of a model enables us to perform repetitive calculations speedily, and we can thus, as in Chapter 7, readily explore variations of model experiments.

These advantages involve some costs: the relationships we deal with may become restricted and rigid, and tend to provide projections rather than perspectives for the future, and, as projections, are little more than extrapolation. More qualitative theories and scenarios, too, emphasise a limited number of variables; although they do allow greater account to be taken of the 'human factor', which may include the idiosyncratic behaviour of individuals or the mass movements of social groups, or changes in values and patterns of behaviour. Scenario building is 'scripted', in that it is more literary than quantitative. In contrast to a computer model, the scenario approach permits a fair degree of flexibility. It is our intention to combine the particular virtues of models and scenarios in our scenario analysis. For the reasons just discussed, we use the models as a basis for understanding particular processes, but the long-run futures we explore are determined primarily by the broader considerations of our scenario analysis.

This leaves us with the question of what to include and what to omit from our computer model. The specializations of research, according to the concerns of policy agencies and academic disciplines, have led to a separation of otherwise intimately-related variables in much modelling. We have, for example, the ironic situation that long-run commodity price projections, such as those by the World Bank (1983), are made almost entirely on the basis of GDP growth rates and income demand elasticities while, in practice, the long-run history of practically all commodities has been bound up with such factors as political turmoil, war, radical technical change, and monopolistic practices.

The aim of our approach is to try to capture socio-political processes in setting up hypotheses about the future, and then to use a global macroeconomic model to assess certain aspects of the evolving scenario. We will set up alternative 'future histories' centred on the assumed implementation of different possible development strategies. (For example, we may follow through the policies and possible contradictions of the Brandt Commission Report, or the implications of strategies such as those suggested in the Lima Declaration.) Obviously, such strategies have to be concerned with the linking of political and economic issues, for example, ways in which transfers of technology and aid or investment may be linked to international treaties and pacts for mutual military or economic security.

The separation of 'political' from 'economic' variables often reflects tradition or convenience. The criteria for separation are also often tautological — economic variables are what we choose to include in an economic model. Nevertheless, *every* variable included in such a model has social and political dimensions. A few examples will clarify this point. At a domestic level, changes in such qualitative factors as 'political alienation' or the 'strength of trade union negotiation' respectively, are likely to mean changes in such quantitative macroeconomic variables as 'labour productivity' and 'income elasticity of labour supply'. A change in the level of social unrest or political instability may well lead to a change in investment or, at the international level, a souring of political relations between certain industrial and developing countries may lead to a slow-down in the rate of technological transfers, foreign investment and arms flows. Again, these examples directly relate to variables in 'economic' models, such as factor productivity or levels of capital stock.

Social and political factors could be included explicitly as equations in a macroeconomic model. An alternative approach is to adjust the conventional economic parameters of a more traditional model exogenously. The choice between these options is a matter of judgement. More adventurous political economists and system analysts have been concerned directly with the formal representation of levels of international tension, or the propensity for social conflict within models (see, for example, Frey, 1978. Guetzkow and Valedez, 1981). In some cases, the model results correspond as closely to historical experiences as do those from models of commodity prices. But other problems face these exploratory studies. There is little consensus about the nature of the relationship between social and political variables: nor is there an accepted common unit of account (such as money) for modelling these issues (with the exception of demography, where people are the units of measurement). In general, the more 'economic' variables included in models are more easily

Figure 1.3: *Links between scenario construction and modelling*

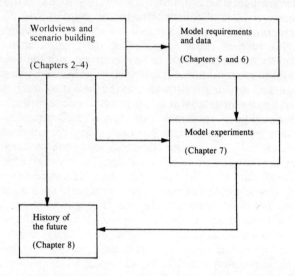

measured and formalised than 'political' variables. One reason for maintaining some separation between 'political' and 'economic' variables may be that the value of the content of an economic model is compromised by the inclusion of more speculative relationships. On balance, for our present purposes, it is more appropriate to include more readily measurable variables into our model, and to deal with other factors in the framework of the more general scenario analysis.

The level of detail (the number of actors and sectors) in the model is not high, which makes it relatively easy to understand, yet it still contains many of the features relevant to the scenario analysis. As with more detailed models, a small number of assumptions determine its characteristic results which are, in the main, unlikely to be greatly affected by the level of detail as such. The model is both an accounting framework and a way of representing the behaviour of actors. For the purpose of conducting experiments with the model, we must adopt a particular set of specified, rigid, behavioural relationships. But for the quantified representation of scenarios we adopt a more flexible approach, which attempts to draw in relevant data from both the qualitative analysis and the model experiments into the construction of the 'future history' to be presented in Chapter 8.

The interrelationships between the scenario building of Chapters 2 – 4, and the modelling exercises of Chapter 5 – 7, are shown in Figure 1.3. They are linked in a number of ways. The model combines a number of key parameters and enables us to examine these factors and their interactions in so far as they can be treated in a quantitative manner. The scenarios provide the context within which particular economic policies may be enacted. Thus, the scenario analysis helps to ensure that the combinations of policies explored, via experiments with the model, are consistent in terms of the worldviews. Finally, the findings of these experiments help us to refine the scenarios when the two exercises are recombined in the last part of the book.

2 Scenario Analysis and Worldviews

It is conventional within futures studies to distinguish between *exploratory* and *normative* approaches to forecasting in general, and to scenario analysis in particular. Normative approaches take as their starting-point ideas of what the future might be like; typically these are based on views of what would constitute a desirable future. They set out to outline the events and processes which would be involved in the realisation of such futures. By contrast, exploratory approaches begin by seeking to establish the consequences of particular trends or events. Thus, they depend more upon judgement of the relative importance of different contingencies, than on choices between different future prospects.

The distinction between the two approaches is far less clear cut in practice than this account might suggest. Elements of value judgement and of technical analysis are required by both. Normative forecasts, for example, have to take account of likely contingencies, otherwise they become utopian fictions instead of casting any light upon matters of choice and policy – although utopias do have their own value, not least as stimuli for forecasting. Thinking about the future, and the role of human action in it, requires an analysis of existing change processes. Choosing which of these processes will be embodied in exploratory forecasts cannot be a purely technical decision; this too requires normative judgements about which issues to focus on, and what treatment to give human agency in the analysis. Thus, exploratory and normative approaches tend to overlap; and, indeed, it is common for one type of forecast to be misrepresented as the other. In particular, visions of desirable or undesirable futures are regularly presented as the result of value-free extrapolations.

A third approach may also be distinguished which synthesises the steps beyond the normative and exploratory approaches. This is the *actor-oriented* approach, which recognises the importance, and the complexity, of the extremely varied human actions that help to

30

determine our future development. This approach, therefore, seeks to identify the major social actors, their orientations towards change, and the conditions under which they might realise their programmes. Scenarios are then constructed which have normative elements deriving from these positions. Different visions of the future are included from the outset, although they are visions that exist in reality as the goals of social groups, rather than as researchers' abstractions. But the scenarios also have exploratory elements. The tendencies which might enable these groups to advance their plans for the future are assessed as part of the forecast.

This actor-oriented approach, known in French as 'la prospective' (Godet, 1978), is most appropriate for long-term social forecasting. We have already pointed to the centrality of power relations in

Figure 2.1: *Scenario analysis*

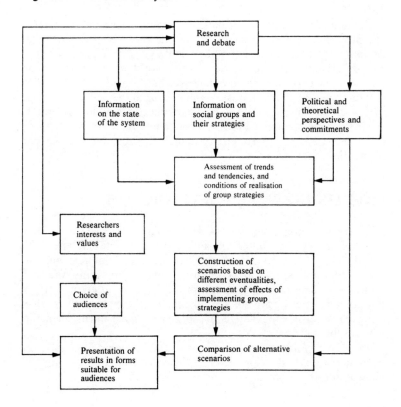

structuring and changing world development patterns. A forecasting method that seeks to address the evolution of these patterns should be capable of dealing with power relations from the start. This is not a matter of discarding the results of forecasting methods like extrapolations; their results can be useful in developing and elaborating the scenario analysis, as long as we are aware that the strategies of different social groups may themselves affect the trends.

Our view of the tasks involved in scenario analysis is schematically represented in Figure 2.1. Some analysis is needed of what processes have produced the current trends and conditions; of the different social groups involved as actors in these processes; and of why these groups have adopted their particular strategies for the future. This involves the use of *theory*, which is fundamental to all futures studies — although this is often obscured by a pretence that the research is value-free and informed by purely technical manipulation of objective data.

A number of basic choices have to be made for scenario analysis. We must decide how to use theory in our analysis, which includes deciding how to take account of the fact that different groups bring to bear different theories and worldviews on world problems. Related to this, we have to decide which groups we shall consider as involved in creating the future; and how we shall approach the task of assessing their various strategies. We also need to choose what information to use in the development of our scenarios. These three tasks are closely interrelated, and we shall approach them by first tackling the question of the different theories.

EXPLAINING WORLD DEVELOPMENT

Theories of development process and strategies are closely associated with the history of social groups and their attempts to influence their world. These theories reflect patterns of social interest and the aspirations and development initiatives that they promote. We shall outline the images of the future associated with different views of development; but we are not arguing that it is the theories themselves that are responsible for change. It is those who make and use them that take on this burden.

In *World Futures: the Great Debate*, we classified different accounts of world development and their associated prescription for change into three broad worldviews. These are:

1. *Conservative*: Best known through the policies of economic liberalism and monetarism, conservative thought draws on, and celebrates, individualist social philosophy. This is seen as promoting the freedom of individual conscience and action, although the need for authority – to prevent the abuse of liberty — is also stressed.
2. *Reformist*: Currently expressed in the economic approaches of neo-Keynesian and structuralism, reformism's intellectual lineage draws on social-democratic ideas of welfare-oriented planning in a mixed economy and pluralist political system.
3. *Radical*: Familiar through varieties of *dependencia* and Marxist analysis of world problems, radical thought stresses the existence of domination and exploitation. Drawing on a range of egalitarian and materialist traditions, however, it argues that this is not an inevitable state of affairs, but that it is reproduced and reinforced by specific social structures and social relations.

The three worldviews have clearly been with us for some time. The formation of industrial societies in Western Europe saw the crystallisation of the first modern forms of conservative and radical thought, with Adam Smith and Karl Marx being towering figures in the growth of these approaches. The worldviews were created out of a momentous change in social affairs, involving a massive restructuring of social relations that encompassed first the West, and finally the whole world. Smith welcomed – not without criticism, for he was no apologist — the emerging capitalist society, which he saw as breaking through restrictive traditional economic and political structures. Some decades later, Marx also saw capitalism as shattering many traditional constraints on progress, but identified it as being based upon the dispossession, exploitation and degradation of workers — they were freed from feudal bonds only to be slaves to the profit motive. Smith's thought was one of the intellectual supports of the new industrial governing classes of what was to become the 'North'; Marx's ideas fused with the aspirations of many members of the new labouring classes created in industrial society. It was not until the second decade of the twentieth century that the first nation chose to take the radical view as the basis for its development philosophy, giving the conflict between worldviews an international, as well as a class, dimension.

Reformism of one sort or another has repeatedly sought to mediate between these two views, and the vast political forces that have been associated with them. Throughout the nineteenth century, there was a

Table 2.1: *Three views of development*

	Conservative	Reformist	Radical
Economic growth	Based on entrepreneurship, flourishing best in free markets with few political interventions.	Most likely to be sustained in mixed economies with government intervention to regulate demand and industrial relations.	Achieved in current profit-dominated system at cost of exploitation and unsocial impacts of technological change: however, also possible under adequate system of planning for social need.
The current economic crisis	Results from 'external shocks' such as oil crises and harvest failures, together with irresponsible government taxation, borrowing and expenditure.	Results from combative pursuit of national self-interest, together with depressed effective demand due to government overreaction to inflationary problems.	Structural crisis of capitalist system, prompted by own internal contradictions and growth of determined opposition in Third World.
International inequalities	Result from failure of nations to specialise according to comparative advantages and to open markets to the stimulus of free trade.	Result from lack of access of Third World countries to capital technology, and influence in international organisations.	Result from a transfer of surplus from poor to rich countries, and subordination of the economies of the poorer to the needs of the latter.
Domestic inequalities	If markets function adequately, inequality result from different contributions of actors to growth, which should itself alleviate inequalities originating outside the market.	Inequality dampens growth by reducing demand and exacerbating political conflict, and requires redistributive policies and more assistance to low-income groups.	Consequence of exploitation of labouring classes by owners of capital and their political associates.

Table 2.1: *Continued*

	Conservative	Reformist	Radical
Preferred global strategies	Closer cooperation between Western countries to revitalise liberal international trade and money policies; reduction of controls on transnational corporations.	International organisations should be restructured to take much more account of Third World interests and need for coordinated global reflation.	Transformation of capitalist world economy by shift to socialist structures; failing that, much greater regional independence and South–South linkage.
Preferred national strategies	Eliminating legislative and pressure group obstacles to factor mobility, technological change and entrepreneurship.	Unless substantial change in world economy, governments should intervene in national markets with sophisticated plans to maintain local control of industry and finance and prevent growth in inequality.	Unless substantial change in world economy, considerable delinking is recommended for poorer countries, who should pursue more self-reliant and needs-oriented strategies.

growth in attempts to mitigate the worst effects of capitalist development by tempering the activity of the market with elements of planning
and philanthropic welfare activities. In the world crisis that spanned
most of the 1920s, 1930s and 1940s, the distinctive intellectual
apparatus of Keynesianism was established, as part of the formation
of new social philosophies and practices appropriate to 'mixed
economies'. It would be too simple to identify this worldview as
belonging to the social classes intermediate between 'workers' and
'owners', but there can be no doubt that the 'middle classes' have
provided a good deal of support for ameliorist and intervationist
policies — at least, as long as their own fundamental interests are not
threatened.

All three worldviews have been elaborated in the most recent
decades of this century, and all are now forced to confront a new
global economic crisis with its political and social ramifications. It is
too early to conclude that any of these approaches will necessarily
gain global dominance. The three worldviews are still recognisable
and distinct, even though each has its own internal debates. (For
example, 'Southern' versions of a worldview are typically very
different from 'Northern' versions, and we could further distinguish
between, say, North American and Western European attitudes.)
The current crisis may well lead to the forging of new intellectual
tools, as we argue in Chapter 8. But, however great a step these
constitute, they are likely to draw on the achievement of existing
worldviews.

Table 2.1 sets out the main characteristics of the three worldviews
in so far as they relate to explanations of economic growth and
stagnation, international and domestic inequalities, and strategies for
change. It cannot hope to capture all the nuances and variants of
each worldview, but it should provide a convenient overview which
can inform the subsequent discussion. Let us consider the three main
views of global problems, and the strategies each is associated with, in
more detail — although we will still only be able to provide the rough
sketches of the main themes of each approach in the available space.

THE CONSERVATIVE WORLDVIEW

Figure 2.2 sets out the main points of the Conservative analysis of the
world's economic problems. This worldview underlies much of the
thinking and policies of Western political leaders and their advisors,
and also of the major private transnational business and financial
organisations. In the late 1970s, it was a common view that the world
economic crisis was basically a result of unfortunate exogenous

'shocks' (e.g. the 1973 oil crisis and harvest failures) and mistaken reactions to these, combined with a widespread trend toward excessive public expenditure. While most western political leaders still strongly support the 'magic of the market', and maintain at least a façade of optimism, more gloomy speculations are now common in many conservative quarters. It has been suggested that the post-war boom was a product of transitory circumstances: these included the reconstruction of war-devasted Europe, providing unsaturated markets for consumer durables; the possibility of transferring workers from low productivity sectors (e.g. agriculture) to work in more productive areas; and a stable set of institutional arrangements, founded on the leading role of the United States. Perhaps all of these circumstances have gone for good, and there may be few satisfactory replacements. Perhaps, too, economic liberty can only be preserved at the cost of some political authoritarianism, to prevent the 'overloading' of open societies by interest groups' demands, the creation of class antagonisms by discontents seeking political power by undemocratic means, and the erosion of public morality by permissiveness. class antagonisms by discontents seeking political power by undemocratic means, and the erosion of public morality by permissiveness.

The problems of restoring high growth rates and injecting some dynamism back into the world economy are still at the centre of the current conservative strategy. Its long-term aim is to promote a liberalisation of the world economy which would enable entrepreneurial initiative to flower. This would be faciliated by a recognition of common interests between Western states and some major Third World countries. They would unite around mutual concerns ranging from the strictly economic, such as free trade and more capital flows, to geopolitical issues, such as taking a stance against 'Soviet designs on the Third World'.

The fulcrum for progressive change in the economy is seen here as the Western world. If steady, non-inflationary growth can be resumed in the OECD countries, then international trade can expand; this will benefit the South, both by providing it with Western markets, and by enabling its integration into international production. This new international division of labour would give Third World countries greater economies of scale and an ability to make use of their comparative advantages. With Western expansion, more aid and investment, especially, can be directed towards the Third World in its export-based growth.

To bring about renewed growth in the West would require both international and domestic change. In terms of the international system this would mean closer cooperation between OECD countries. As in the immediate post-war period, some means of establishing

Figure 2.2: *The conservative view of global problems*

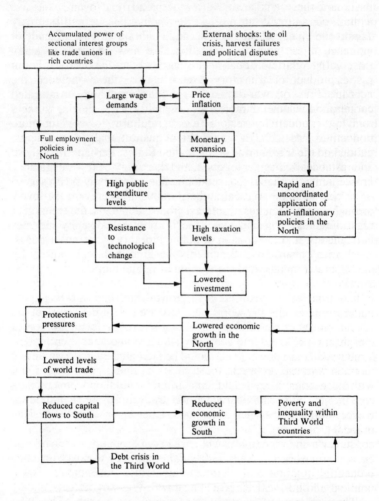

adherence to a set of common economic ground rules is required. This is because, although such ground rules are seen as being in all countries' long-term interests, there are bound to be occasions when short-term interests, pressed upon governments seeking re-election, may deflect the long-term aims of individual countries. A reinvigorated United States might assume the role of leadership that the West requires to establish a new international regime. Or is it possible that the ground rules could be worked out by collective agreement between a small number of leading countries — such as the United States, Japan and West Germany — which would be in a position to influence other OECD members? Such cooperation was proposed in the 1970s by the Trilateral Commission (who were too closely associated with certain reformist currents to be popular with many hardline conservatives) and other international bodies also urge cooperation within the North.

Whether under United States leadership or not, Northern cooperation would help to coordinate demand management, energy conservation and other national economic policies, so as to promote growth and price stability. Crucially, it should stave off the threat of protectionism, and present a united front against Soviet expansionism or 'blackmail' by Third World producer cartels. Consultations through existing organisations, it is hoped, would be able to accomplish these ends; there is little need to add to existing international bureaucracies.

Growth in the OECD area would be organised around technological change. While earlier economic growth involved extensive development drawing on pools of migrant, female and displaced agricultural labour, there would now be a growth in productivity, particularly in the service sector, through a 'new technological revolution'. Microelectronics technology (and perhaps biotechnology) can permit rapid productivity increases, and may create a whole range of new products for which vast demand may be generated, according to conservatives. Although some unemployment will be associated with these technologies, new industries and jobs should emerge. Full-employment policies contribute significantly to high levels of inflation by permitting excessive wage demands and disruptive working practices, and market forces must be allowed freer play in regulating consumer demand, employment and wages. There are no inevitable class conflicts, and with greater economic freedom people with entrepreneurial qualities will be able to achieve social mobility, while political and business incompetence will rapidly be detected.

The problems of adjustment in the rich countries need to be tackled cooperatively, so as to limit the social and political cleavages resulting from these changes, and face governments with less temptation to abandon international agreements. But what is required above all, according to conservatives, is government determination to resist and reduce the excessive power of trade unions and other sectional groups, and to enhance individual economic liberty.

Since the improvement of the world economy is seen as being dependent upon Western recovery, the industrial countries need to convince the South to avoid actions such as increased oil prices or the escalation of political conflicts, that could disrupt this process. The North should encourage export-oriented growth policies in the South by active measures of trade liberalisation. Liberalisation can proceed by steps, with reciprocal agreements or, better still, multilateral restraints on protection. It would probably need a supportive system of compensation for the costs of adjustment, which might also be available to industrial countries facing Third World competition. Short-term political pressures in favour of protection could thus be resisted, and a liberal international order constructed.

The South, then, is seen to benefit from Northern growth; it can achieve greater access to expanding markets, and gain from a 'trickle-down' of the world's wealth and the new technologies (which could be especially useful in their export-oriented sectors). The newly-indust-rialising countries in particular would absorb the new technologies (this would require some reform of domestic institutions) and continue to move toward more capital-intensive, quality manufactur-ing. Other Third World countries could make use of their comparative advantages to create resource or labour-intensive exports. The NICs might provide capital for the development of these other countries, if they can overcome their own debt problems (which will probably mean periods of greater austerity for them).

Existing international institutions — such as GATT, the IMF and the IBRD (World Bank) — would play an important role in enabling these developments to proceed smoothly. In general, these institu-tions should aim to operate by promoting consent. They should encourage market-based adjustments — although these will not always be possible given the uneven distribution of resources and the inertia of vested interests. With the recent unhappy experience of inflation, debt and balance of payment crises, international finance needs special attention. The over-rapid growth of private credit for the South is a source of instability and some conservatives hope that official development assistance will be increased. With more official loans and, as other conservatives emphasise, more direct foreign investment (including, perhaps, an international investment trust), the need for private lending may decrease. However, many conservatives argue that banks should by now have learnt their lessons, and can in future be more reliable and foresighted sources of loans than government bureaucracies could ever be. Since fluctuations in trade and commodity prices are a major source of liquidity problems for the

Third World, some system of compensatory financing may be desirable, although again there is disagreement among conservatives here. As in all bureaucracies, there is the possibility of abuse of such intergovernmental structures, and funding bodies would need to be able to scrutinize their activities closely.

To summarise, the image of a desirable future proposed by the conservatives is one of a new international division of labour, based upon dynamic technological change in the North. All parties would gain from increased trade. Political interventions should be designed so as to facilitate, rather than subvert, the optimal allocation of productive resources through market processes. Implicitly, governments will take more of a back seat: their role is to provide the conditions in which transnational corporations and local entrepreneurs can successfully upgrade technologies and economies, thus promoting growth. Growth does not necessarily mean global or national equality — indeed it is likely to be hindered by redistributive policies that deter innovators and investors. Achieving a distribution of income and wealth that reflects ability and effort, rather than traditional privileges or bureaucratic dictates, is seen by conservatives as only following from this sort of strategy.

THE REFORMIST WORLDVIEW

At their bluntest, *reformists* accuse the conservative approach to international development of being a vision of the world weighted in the interests of the North. Since a relatively liberal system did little for the South during the post-war boom, why should it accomplish more now? Why should the South postpone its claim for global equity, in order to strengthen the OECD economies? The world, it is argued, is profoundly divided between quite divergent types of economies, and abstract notions of 'comparative advantage' and the 'gains of trade' bypass the real realities of cumulative disadvantage in an unequal world.

A major forum in which attempts were made to account for the failures of liberal economic policies was the United Nations Commission on Trade and Development. In UNCTAD 1 in 1964, Secretary General Prebisch argued that liberal doctrines ignored the significant structural differences between what he termed the 'industrial centres' and the 'peripheral countries'. Even institutions like GATT operated in line with the interests of only one type of country — GATT effectively excluded, for example, many products of interest to Third World countries from its tariff reductions. The deviation of the West from liberalism, when this threatens its vested interests, only compounds the inequities that have been established.

In this analysis, one of the chief structural differences between the centre and the periphery is that the former specialises in capital-intensive manufacturing and the latter in labour-intensive primary production. Increases in world trade have tended to benefit the former. The income elasticity of demand for manufactures has been higher, while the terms of trade for primary products – and thus the relative economic power of their producers – are liable to decline, because technical progress lowers their prices more than those of

Figure 2.3: *Reformist view of world problems*

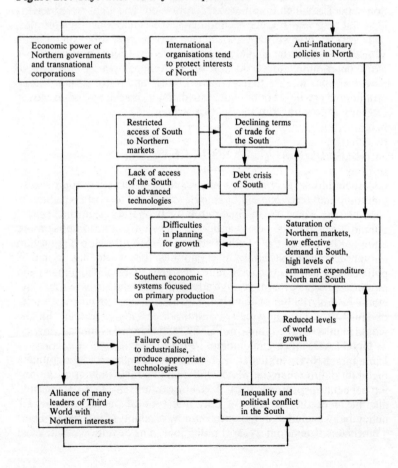

industrial goods. The Third World's ability to purchase goods falls, and it suffers balance of payment crises, under-capitalisation, and related problems.

Figure 2.3 schematically represents the reformist analysis of the present economic problems of the world, which are seen to be deeply embedded within its economic structure (compare Figure 2.2), and to require the reform of that structure. It is assumed that appropriate reforms could promote equity within the trading and financial systems, restoring world economic growth and raising productive capacity and levels of consumption in the South. Many of the elements of such a reform are seen as being contained within such international proclamations as the Lima Declaration, and the resolution of the United Nations 6th Special Session. But these have had far less practical implementation than was anticipated.

In essence, the strategy proposed here is one in which the Third World persuades the richer countries to agree to a reorientation of world production and technology. Since the special difficulties of the South are, in large part, related to its specialisation in primary production, its rapid industrialisation is seen as central. The Lima Declaration, for example, set a target that 25 per cent of world manufacturing should be based in the Third World by the end of the twentieth century. This would be achieved in various ways: funding to encourage industrialisation (with private investment subject to more scrutiny); general funds to help offset the difficulties of trade for the poorer countries; greater access of Southern manufactures to Northern markets; and considerable technology transfer (under appropriate supervision). Third World countries would then become much freer to pursue strategies of industrialisation, essentially along the paths already followed by the North. Global détente and reduced levels of military expenditure would facilitate such developments, and the policies set out above would need to be elaborated, implemented and monitored by restructured international organisations. The North would necessarily have to accept a considerably increased proportion of Third World produce in its consumption of manufactures, which would in turn require considerable restructuring of its own industries.

UNCTAD's own calculations (1981) also showed that, for the Lima targets to be fully met, the South would need an exceptionally high rate of growth in both GDP and manufacturing output (7.5 and 9.6 per cent, respectively). The South's output of capital goods would also be increased rapidly, because the strategy implies a focus on heavy industry, with an integration of the technological structure of Third World economies, and a rapid upgrading of relevant skills. In turn, this implies considerable resource transfers, possibly with considerable expropriation of foreign firms who fail to fall into line promptly with national development priorities.

These major changes in Third World production carry considerable consequences for world trade. Two vital issues concern market access and price stability (although others arise around points like the control of world shipping). Increased Southern access to Northern markets means reduced tariff and non-tariff barriers to the South, especially for critical primary products. Prices should be stablilised to maintain at least the South's export earnings, with insulation against the effects of Northern inflation. To this end, buffer stocks, indexation and methods of compensatory finance – including softer loans — are called for. Negotiations around such issues should be multi-lateral, and result in an expansion and diversification of Third World trade.

Clearly, all such proposals require considerable finance of one sort or another, as does any approach to resolving the existing debt problems of the South. To these may be added increased aid levels in general (the target of 1 per cent of rich country GNP is often cited), special loan and sales terms to the poorest countries, encouragement for divestment, and greater funds for development agencies and banks. Many reformists call for a Marshall Aid-type plan to stimulate the rapid expansion of Third World markets. In addition to aid and public transfers, money for such schemes might be produced by sharing the profits of common resources (e.g. the oceans and outer space). Considerable resources could be released by redirecting armaments expenditure, which currently forms a percentage of the world product equal to the total income of the poorest 15000 million people. To ensure that funds are managed in Third World interests, and are less subject to geopolitical considerations, schemes should be made more automatic. Agencies should have greater participation on the part of Southern countries, and new Third World-oriented bodies might well be set up.

As well as the conservative worldview being weighted in favour of the North where international affairs are concerned, it is seen by reformists as being weighted in favour of privileged élites in domestic affairs. For example, the pursuit of monetary and deflationary policies may curb inflation, but it is done at huge cost to the more vulnerable social groups — those particularly affected by unemploy-ment and reduced welfare outputs. It is such circumstances that breed class conflict — just as international inequalities breed international conflict — when what is needed for economic and social regeneration is cooperation. This can only be based on a recognition of the different circumstances of different social groups — states should compensate for inequalities of opportunity, so as to ensure democracy and ward off social tension. This does not involve aiming for complete equality, but continuing redistribution should eradicate 'cultures of poverty'.

To summarise the reformists' image of a desirable future, international economic expansion would be spearheaded by a steady redistribution of world manufacturing production toward the South, which experiences rapid growth and industrialisation. Northern countries would find increased markets in the Third World, and many reformists stress this as the solution to the world economic crisis (which they see as related to a lack of effective demand).

However, the North's exports of heavy products would fall and imports of manufactures rise, in relative terms, under such a strategy. This might require the industrial countries to change their consumption patterns and life styles. Such changes might well be necessitated in any case, by the mounting social crises within the West, which reflect the problems of its present way of life. These large changes may involve considerable short-term costs, for at least some groups in the North. For this reason it is likely that many Northern reformists, in practice, hope at best for only a partial realisation of the reformist policies outlined here.) Internal socio-economic changes would clearly result in the Third World too. These would be related to the expansion of the industrial working class, of scientific and technical workers, and of new urban centres; this should lead to improved income distribution. However, to facilitate this, deliberate action by government must also be encouraged, especially to erode the traditional interrelations of poverty and privilege.

THE RADICAL WORLDVIEW

Conservatives challenge the reformist approach outlined above, on the grounds that its massive costs will lead to an inefficient allocation of resources, shackle enterprise, and cause inflation. Drastic and rapid changes in the ways of life in the North and South would lead to political instability and create the conditions for totalitarianism. Global regulatory agencies would be prone to corruption and politicisation, undermining their authority. The burden of public expenditure would be a further source of disruption. And the centrally-planned economies could pursue their own global designs with much less hindrance.

The *radical* worldview shares some of these criticisms of reformism, while dissenting widely from others. The radicals do share the reformists' view that free markets often amplify existing inequalities with particularly destructive results for poorer countries and vulnerable social groups. It is not seen as likely, however, that the industrial countries would voluntarily acquiesce to a restructuring of the world order, and of their own ways of life, in the interests of the Third World with particularly destructive results fro poorer countries and vulnerable social groups. It is not seen as likely, however, that the industrial countries would voluntarily acquiesce to a restructuring of the world order, and of their own way of life, in the interests of the Third World (at least, not without their having already undergone huge political changes). Even if the North formally accepted new institutional arrangements, the interests of its dominant groups would continue to be expressed to the disadvantage of the South. Consumer preferences, new technologies, and educational and cultural practices would still be dominated by the richer countries, and thus world development would continue to depend on the internal dynamics of these societies.

This third approach, then, sees the linkages between the North and South as extremely deleterious to the latter. No amount of minor adjustment with these linkages would enable the South even to begin to achieve styles of economic development that significantly benefit a great proportion of its population. Certainly, painlessly negotiations between the two unequal partner regions can accomplish this. Confrontation is inevitable. The problem of the Third World involves economic, technologicial and political dependency; and increased North–South trade and financial flows would be likely to reinforce this in subtle ways.

Figure 2.4 depicts the radical view of international affairs, which portrays the contemporary world economy as reflecting class domination. Most fundamentally, there is conflict — often disguised or displaced — between classes whose labour produces the economic output, and those that control the production and consumption of the surplus product. But the dominant classes in the Third World countries are, in the main, effectively subordinated to those of the North, although they are rewarded by substantially higher levels of income than those of other social groups. This is one reason why the Third World has relatively high internal inequality, and fails to pursue the path of national development that industrial countries have taken. Another reason is the lack of its own Third World to colonise and exploit. A less dependent, less imperialist world system is required, which means changing the balance of class forces in peripheral countries, so as to embody the interests of peasants and workers in government policies. This would naturally be helped by changes in power relations in the industrialised countries.

Figure 2.4: *The radical view of world problems*

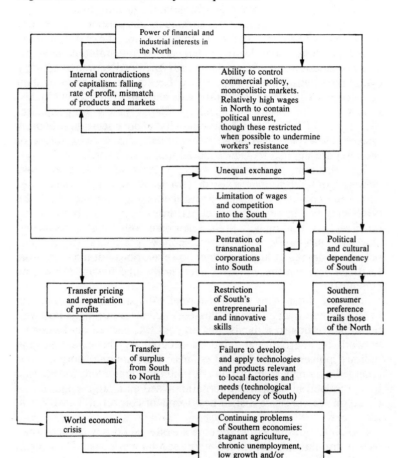

export possibilities than by domestic consumption, and are dependent on core country suppliers for capital goods and associated technologies.

Advanced technologies are produced and employed in the Northern countries where industrial revolutions were typically accomplished, or well underway, in the nineteenth century, and where productivity is now relatively high, and wage labour is the norm. The radical worldview stresses that these core economies are internally integrated:

strong organisational and infrastructural links exist between the
sector that produces mass goods, and the sector that produces
consumption goods for the domestic market. The peripheral countries,
in a world that has been shaped by the often violent expansion of the
core economies, have disarticulated, poorly integrated economies.
Their production and infrastructure are typically determined more by
export possibilities than by domestic consumption, and dependent on
core country suppliers for capital goods and associated technologies.
Levels of productivity are usually relatively low in peripheral
countries, but even in high productivity export sectors, labour is
typically low paid and often overworked, and various forms of
coercion are applied to secure labour and limit costs.

The position of countries within the world system as *core,
periphery* or *semi-periphery* also determines the social class structures
within the country. (Semi-peripheral economies, among which the
NICs may be classified, in effect perform a buffer role between the
other regions, trading with both. Their own political alignments and
class structures reflect their former core or peripheral status.) The
balance of forces at any one time, however, does not mean that the
relations between these classes necessarily tend toward a foregone
conclusion.

The core states of the world historically shaped the world system in
favour of their own dominant classes' interests. During the post-war
boom, according to radicals, this process was carried out through a
network of international relations secured by the economic and
military power of the United States. Transnational corporations
emerged as a new form of powerful international actor; for the first
time, production itself was internationalised on a large scale. Firms
based on the core established new forms of control of Third World
resources.

Dominant groups within the core are often at odds with each other,
as was dramatically evidenced by the second world war. The general
benefits of the post-war order to the core were at the cost of a steady
economic decline of the USA relative to other western powers, which
has led to new conflicts within the core. Economic competition
intensified among core nations, and core interests are challenged by
new forces in the world. Global conflicts have beome more visible,
with a political decline of United States influence *vis-à-vis* the
socialist bloc and Third World liberation movements. The revival of
the 'cold war' is seen by radicals as an attempt to exert leadership and
achieve western unity through political–military power in the face of
this challenge. Within the core countries, the crisis is viewed as a

product of the very success of the post-war boom. On the one hand, technological changes and international competition are throwing whole industrial sectors into turmoil; conversely, the social institutions that secured poltical consensus have become too expensive a burden for private capital. The post-war boom was based on a suppression and displacement of conflict rather than on cooperation — now the conflicts have returned, in a variety of guises.

The programme for change that stems from the radical worldview requires a transformation of the relationships of dependency. Short of a coordinated restructuring of the world economy, which seems extremely unlikely, Third World countries need to break away from the sources of their problems, to cut their links with the North selectively. Major problems would be faced for most individual countries seeking self-reliance, but a pooling of financial, human and natural resources would clearly provide a much more adequate basis for autonomous development. Thus, radicals are interested in possibilities for economic and technological cooperation among developing countries (ECDC and TCDC) and aim to capitalise on, and go beyond, those limited gains that might be won from reformist strategies.

A considerable increase in South–South trade, free from Northern intermediaries should be organised. Third World finance and transport systems would be required. Substantial government intervention is needed for such restructuring; beyond the integration of markets more integration of economic sectors is required, particularly where industrialisation is involved. Industries, especially manufacturing of heavy machines and infrastructure, will often require planning at a regional level. This can prevent unnecessary duplication of effort, achieve substantial economies of scale, and help obtain complementary development profiles. Agricultural production, and associated activities such as fertilizer manufacture, also need to be expanded, with the aim of rapidly relieving the dependence of many Third World countries on food and food aid from the West.

Within this general framework, trade between Third World countries would take place with considerably lowered tariff barriers, and with such instruments of trade support as state trading associations, export credits and international clearing systems. These efforts should be accompanied by programmes of scientific and technological collaboration, shared educational facilities, and the development of Third World-based mass media, as measures to counter inappropriate external cultural influences. Among Third World countries, technological information would be available on advantageous terms to shift the direction of technological change in the interests of the South. In this way, the new technologies could make the maximum use of and help to further develop, local skills and knowledge. They would involve labour-intensive techniques, relying much more on local resources, and industrialisation more supportive of agricultural development. Southern consumption would be reoriented to more indigenous life-styles, away from a dependence on western culture, and with much greater equality and satisfaction of basic needs.

These strategies call for financial resources and the involvement of international organisations. Some Northern countries might be sympathetic to these changes, but most would need to be persuaded to support (or at least not impede) them by collective pressures, especially the use of raw materials as trade weapons. Those North–South transfers that were available would be managed rather as in the reformist strategy, but there would be much more concern with the local administration of funds and the regulation of transnational corporations and Northern agencies of all kinds. Third World-based development banks, monetary systems and clearing unions, with local credit facilities, might be created; one source of funding here might be revenues based on oil exports. Further to such specialised institutions — to which might be added scientific and technological associations, and producer/consumer trade associations — a general reorganisation of United Nations agencies to serve Third World interests is needed.

Since the radical view of North–South relations implies that the periphery is exploited by the core through trade and finance, a change in the balance of these relations should provide more resources for local development needs. At the present time, trade functions as the medium for unequal exchange, and one instrument proposed for reforming it is a Southern customs union. This would tax North–South trade to achieve just returns on exports, to discourage imports, and to allocate the funds received to the new agencies discussed above. Producer cartels formed by the South would compensate other countries in the union for price rises and seek to maintain prices advantageous to the Third World as a whole.

These proposals for a new set of principles require considerable cooperation and support from a large number of peripheral countries. This unity would need to be strong enough to resist inevitable overt and covert opposition from the West, and attempts by the Second World to use the situation to their narrow advantage in East–West relations. Substantial changes in many countries are required, which may well take considerable time to achieve, with many setbacks on the way. For single countries to pursue radical strategies without support is very difficult: they would have to accommodate to the international situation, while achieving as much selective delinking and self-reliance as possible. This may be the only choice in the near future — especially if the 'oil weapon' loses its potency in the face of a restructuring of world energy demand.

To summarise then, a *radical* version of the future presents a picture of a complete break with past patterns of dependency and injustice. Global economic growth may not be easy to achieve, since it would demand a drastic restructuring of the world economy. Entrepreneurial and other skills in the Third World would be reoriented towards production for local needs. There would be massive efforts to raise the health and education levels of the poorest social groups; the pattern of alliances in the world would reflect the emergence of these social actors on the stage of world history.

Proponents of the other worldviews see radical proposals as being unrealistic, and as threatening to plunge the world into anarchy or totalitarianism. Radicals argue that the oppressed can, and will, fight for their own freedom, and cannot depend upon others, or any impersonal economic forces to do this for them.

CONCLUSION

Chapter 1 sketched in some of the dimensions of the current world crisis. We have now seen that different worldviews explain these phenomena in different ways, pointing to distinctive underlying processes and divergent strategies for change. Our scenario analysis will take these worldviews as the intellectual underpinning of the strategies of different interest groups concerning world development. We shall apply this approach to explicate prospects for the future.

3 Global Strategies and Scenarios

Each of three worldviews outlined in Chapter 2 presents distinctive ideas of a desirable world order. They suggest different strategies for governments to pursue, both in the event of progress being made towards the preferred international system, and otherwise. However, the prescriptions offered by each worldview do not necessarily exhaust all the possible futures that they can envisage. Continued economic crises without distinctive resolution form one such possibility. Even more ominous is the prospect of a major international conflict, with weapons of mass destruction threatening our planet's survival. This chapter will demonstrate how scenario analysis may tackle these issues.

WORLDVIEWS AND GLOBAL SCENARIOS

We begin our scenario analysis by thinking about prospects for the world system. (The next chapter will take up the circumstances and strategies of the partiuclar types of country that are related in this system.) We shall consider, first, the preferred changes in the international order suggested by each worldview; second, the conditions under which these changes might be realised; and third, possible futures resulting from the failure to satisfy any of these conditions. To begin then, what global futures are seen as desirable by each of the three worldviews?

The conservative worldview supports the removal of barriers that limit the mobility of capital, talent and ideas; it proposes the consolidation of a liberal economic order as optimal for world development. This means reversing trends towards protectionism, and controlling inflation and re-establishing confidence in the financial system so as to limit speculation and unwise borrowing. Under these circumstances, rapid economic growth in the industrial countries will 'trickle down' through the world economy for the benefit of all. Such goals are proposed by the leaders of many western countries, by international associations like the OECD, and by agencies like GATT concerned with maintaining the structure of the post-war settlement.

There is disagreement in the West over some aspects of this type of policy, especially about the degree to which costs of change should be shared among countries. Western governments, and Western-oriented Third World regimes, face very different magnitudes of, for example, structural unemployment, energy dependency and military expenditure. Thus the degree to which the initiative for change is the United States' responsibility, and how much it should be required to forsake its internal interests in order to stimulate world reflation and restructure international agreements, is a matter of contention. Even within the NATO group of countries, the degree to which the state of East–West relations should be the critical arbiter of trade with the CMEA bloc and support for Third World regimes, and whether rearmament should be given priority over productive economic investment, are hotly contested.

In the rather muted North-South dialogue that followed the United Nations 6th Special Session, the conservative strategy was opposed by Third World governments arguing for a reformist strategy. The positions put forward by international representatives of Third World interests (such as the Group of 77 and the UNCTAD secretariat) call for a reform of the world economy so as to distribute its wealth rather more evenly. The present world crisis should be taken as an opportunity for a restructuring to reduce global inequalities as well as to restore stable economic growth. Resource transfers to the Third World, with greater access to the markets and technologies of industrial countries, are the main planks of this programme; and sympathisers in the West argue that such a New International Economic Order (NIEO) could be beneficial to the rich countries, too, providing new outlets for their production and capital.

This strategy fits the account of the reformist worldview in Chapter 2. The force behind the call for a NIEO has declined since the mid-1970s, with little sign of the Third World achieving significant concessions from the North. In part, this reflects problems in establishing Southern unity. Various proposals for South–South cooperation have made less progress than anticipated: countries often compete for favours from the North and regard each other with suspicion and hostility.

The *radical worldview* also faces the problem of a lack of solidarity among Third World countries. It argues that any restructuring to improve the economic prospects of the richer countries is likely to reproduce the dependency and subordination of the South. Instead, restructuring of international relations for Third World countries should involve a major delinking of existing economic and political ties, with selective relinking, between Third World countries especially. While the consequences for the industial world are uncertain, proponents of selective relinking argue that, only by overcoming the dependency inherent in existing global relations, can the poorer countries successfully overcome their problems in the 1980s and beyond. It would be desirable for the North to participate in a real change in international relations but, at present, there is more prospect of the peoples of the South seeking autonomy than there is of the North instituting sociopolitical changes that would break the chains of dependency.

Some support for Southern radicalism might be forthcoming from progressive European movements, but there are also dangers of being drawn into the East–West conflict. One of the dilemmas faced by the radical worldview concerns how far cooperation can be achieved between groups in the South, and groups in a divided North, in the restructuring of international arrangements; and how far it will be possible for Southern interests to be imposed on current arrangements, given the risk of retaliation or political or military intervention. The extent to which cooperation between Third World countries can be focused around collective self-reliance in a transformed international order, rather than around improved status in a NIEO, depends on the future leadership of the Third World.

The three worldviews are mutually critical of each others' strategies, suggesting that they would lead to extremely different futures from those implied above. For example, conservatives argue that the reformists' and radicals' desire for the regulation or displacement of market processes can lead to bureaucratic bungling, to discourage-ment of entrepreneurs faced with continual prospects of expropriation and political redirection, and to the loss of managerial and accounting skills accumulated in private enterprises. A politicisation of economic affairs is a likely outcome; decisions would be taken in line with political manoeuvreing, leading to a dangerous neglect of sound economic criteria in order to satisfy sectional interests. The restruc-turing proposed by reformists and radicals could easily fail badly, leaving the world with enormous obstacles to restoring trustful, prosperous international exchange.

This critique raises important questions about the conditions for change, and the ways in which economic gains and political power would be distributed in a restructured world. But the conservative strategy itself may be criticised in similar terms. Reformists and radicals would argue that its assumptions of free political consent to, and equitable distribution from, the operations of a liberalised world market, are exceedingly precarious. Internal pressures in the North hardly seem to be conducive to industrial countries spearheading a trade liberalisation of the order required. The costs of adjustment in this strategy seem excessively high, especially considering the likely maldistribution of gains. A substantial measure of political authoritarianism seems to be implied, at present, by strategies of economic liberalism.

The radical strategy would be criticised by reformists on the grounds that it calls for tremendous recomposition of the political leadership and orientations in the South to bring about the sorts of solidarity and realignment suggested. How could the cultural and business links of many important Third World élites, and the established consumption patterns of the masses, be changed so drastically – short of draconian measures? Radicals would challenge reformists to provide substantial evidence for the supposed commonality of interest between North and South. If this exists, why is the North so resistant to a NIEO? And who within the South would gain from this strategy — the masses, or the élite groups benefitting from trade and other links with the North?

The conditions for realisation of the desirable futures of the three worldviews depend on certain fundamental sets of relationships in the international system. The main regional relations pointed to in the above discussion are: North–South relations; North–North relations (subdivided into East–West relations and West–West relations); and South–South relations. The degree of cooperation or confrontation *between* regions, and the degree of unity or disunity *within* regions, are identified as key issues. The content of relations between regions will depend on the dominant strategic posture and the internal relations of the regions concerned. We do not include East–East relations within the schema above since, despite the events in Poland in the early 1980s and other signs of stress in the Soviet bloc, we follow most other commentators in seeing this region as being far more under the hegemonic discipline of its superpower than is the case for the West. (The People's Republic of China we consider as part of the South.)

Let us consider the conditions for realisation of each of the three strategies. To do this means stepping down from our Olympian perspective and making some commitments and choices related to the three worldviews. While drawing on the insights and the mutual critiques of each worldview, the material to be presented reflects rather more of the radical worldview, since its attempts to grapple with unequally distributed power, and to ground human action and ideas in these power relations, are particularly valuable in understanding the course of world development. It is also concerned with accounting for the problematic record of those movements that have sought to follow its prescriptions in the past, although conservative and reformists have been consistent in their critique of the limitations of planning and the dangers of bureaucratic power.

Tables 3.1–3.4 present information on the factors that could promote different patterns of change in the key relations identified earlier. While we have drawn mostly from the world-system and political economy literature in identifying these factors, so that the balance given to them might be considered inappropriate by some readers, we hope that they will be recognisable by proponents of different worldviews. These tables necessarily contain cross-references to each other. Each contains a range of factors that might promote the same end, and there are innumerable ways in which they might combine to reinforce or subvert each other. Relationships within the North and South — themselves very heterogeneous groups — are seen as vital to the course of North–South relations. We can formulate scenarios drawing upon the factors cited, so as to provide 'future histories' of how each of the worldviews' desirable futures might be realised.

In very broad terms, and with no attempt to indicate a timescale for the hypothetical events, we can summarise these three scenarios as follows: the desirable conservative future of a *liberal international order* would be most likely to result from West–West unity, low East–West conflict, little South–South unity, and Northern strength in North–South relations. The second worldview, the reformists' *reformed international order*, would result from West–West disunity, low East-West conflict, but considerable South–South unity, promoting Southern strength in North–South relations. And the radical scenario of *collective self-reliance* would imply no more than moderate West–West unity, possibly heightened East–West conflict, high South–South unity, but relatively little opportunity for this to be expressed in North–South negotiations.

Table 3.1: *West–West relations: factors affecting the unity/disunity of Western industrial countries*

(a) Geopolitical factors

Unity in international affairs is more likely to be obtained if one country is able to exercise strong leadership and thereby set the framework for negotiation and compromise.

In the event of sustained mounting East–West conflict, Western nations might tend to accept US economic leadership for military and political support: the same might be true of North–South confrontations or serious instability within the OECD group. (Such realignments within the West could be more or less stable, with the timespan being very important here.) Other types of global conflict, e.g. Sino-Soviet or intra-Third World, could lead to more or less incentive to impose a more unified organisation on Northern affairs, depending on the ties of Western countries to the conflicting countries. Other 'shocks' like climatic change or major technological innovations, could pose strains for Western unity, especially if partners are affected in different ways. The relative positions of different Western countries would be affected by the degree of control available over markets, investment zones and resource supplies. Countries with more access to these would be advantaged economically, and thus the ease of retaining or challenging a position of dominance in the West would be increased. Shortage or high prices of energy and resources for example would advantage certain Northern countries: the US could cope better with high prices on account of its considerable indigenous reserves, given conservation policies. Political prospects, and adequate planning where nuclear power and coal mining are involved could be vital.

(b) Political orientations of North

Divergent types of government within the West, pursuing different domestic and foreign policies, would have difficulty cooperating where North–South relations are concerned. Political changes within Northern countries might be managed so as to permit more or less coincidence of interests, but there has been a tendency for groups of Western countries to stay out of step over the last few decades. In part these political

differences relate to the various industrial structures of Western countries, and it may well be that differences in the ability to innovate and apply new technologies, and to engage in the economic restructuring required to retain market competitiveness will affect the convergence/divergence of the North. In principle, the greater size and global reach of the US should favour its restructuring, but institutional and political obstacles remain. Political stability in the US as compared to other Western regions — in the former conflicts might mainly revolve around wage levels, in Europe and Japan around welfare policies and employment protection respectively — might favour a US leadership of the West. European disunity would reduce the tendency toward Europeanism by inhibiting mergers, economies of scale and military cooperation within the EEC.

(c) Economic policies

Drift toward protection could easily provoke a chain of retaliations which would hamper Northern unity. The lead given by technologically advanced industries (typically favouring liberalisation). and the mobilisation of regionalist sentiment and declining industries (where the unions may ally with capital in seeking government support) around protection versus liberalisation is important. The role taken by more dynamic countries in international organisations is very significant. Competition over resources and markets can grow apace during recessionary periods. Policies concerning access to resources whether those obtained by trade or direct extraction (e.g. ocean resources), and markets, and concerning the regulation of activities, may be difficult to forge; but the risk of losing other Western markets can facilitate compromise here. The use of trade as an economic weapon (e.g. in East–West relations) may continue to impede Western unity.

Table 3.2: *South–South relations: factors affecting levels of unity and cooperation among Southern countries*

(a) Increased Third World power

There are several routes to increased bargaining power for Third World countries in North–South relations, and some of these would interact positively, being reinforced by and reinforcing unity. A failure of industrial countries to free themselves from reliance on energy and raw materials imports — for example, through failures in energy programmes — would enable Southern demands for reform (or non-intervention) to be tied to the availability of resource supplies. Successful experience in negotiations could encourage further collaborative efforts and enable principles and mechanisms of cooperation and regulation to be established. Northern disunity or economic crises would increase the relative power of the South, as would the siding of some industrial countries (or significant groups within the North) with Third World demands. The failure of the economic growth models of the industrial world — and its client states — might encourage the search for alternatives.

(b) World economic tendencies

Trends towards protectionism in industrial countries, especially against Third World manufactures, could sharpen global polarisation and increase the felt need for attempts to define Southern futures. Conflicts within the North might provide more room for manoeuvre and bargaining, or might take the form of attempts to establish 'zones of influence' divisive of Southern countries. Northern countries could also act to divide the South by selective protectionist measures, increasing competition among Third World countries to establish themselves as suppliers to the North. Leadership positions might be granted to wealthier countries in exchange for guaranteed access to financial and national resources: particularly if, say, OPEC countries were prepared to exercise producer power in international negotiations. A strong blend of inducements and sanctions would probably be required for any *single* country to assume a significant hegemonic role. Cooperative leadership by a group of powerful Third World nations would require a degree of political unanimity that would probably have to extend to the sphere of domestic politics as well as global strategies. If these exchanges reinforced or amplified existing tendencies, the result might be centrifugal in terms of Third World interrelations. This would be particularly likely if such difficulties coincided with serious conflict, and would be less likely to the extent that institutional flexibility had been established through prior regional cooperation, so that diverse Third World interests could be more readily accommodated.

Table 3.2: *Continued*

A Third World interest coalition would be inhibited by Northern negotiation and action in economic and political affairs which tended to reinforce existing differences within the Third World. Two such differences in particular are those based on different levels of economic development, and on geography. In the former case, the North may establish distinct types of agreement with wealthier Southern countries, creating 'buffer zones' and sub-imperialisms' to blunt North–South conflict; in the latter case regional 'zones of influence' may be established by different industrial regions in a multipolar world. In either case relationships within the Third World would be fragmented, and the emergence of leadership within the South rendered more problematic.

(c) Political development

The heterogeneity of the Third World suggests that convergence in foreign policies may reflect either global polarisation or internal political change, but not economic convergence. The victory of nationalist and anti-imperialist movements would have to rapidly crystalise around similar social groups (class interests) in order for much unity to be developed. The development of cultural exchange, or mass media capable of promoting Southern viewpoints and the like would help forge this, although there is considerable potential for radical nationalisms to lead to conflict among neighbouring states with conflicting cultural traditions.

(d) 'Role of semiperiphery

The 'middle-income' Third World countries, particularly the newly-industrialising and resource rich countries, may play a key role in determining the prospects for Southern solidarity: structurally, they may be integrated more closely with the North, or ally more with the poorer countries. They are in relatively favoured positions for exercising leadership in the Third World, although fears of regional imperialism or sub-imperialism may have a real grounding.

59

Table 3.3: *East–West relations: factors influencing levels of conflict and coexistence between East and West*

(a) Economic factors

Economic difficulties within either bloc may create internal strains which may (i) induce states to use external threats as a basis for reducing domestic unrest or marshalling allies into cooperation, at the cost of worsening East–West relations; (ii) increase pressures from strategic but declining heavy industries for military contracts, which lead to increased arms races and have to be legitimised in terms of potential conflict; and (iii) conversely, create opposition to military budgets from other sectors of the economy faced by decreased state purchases and from welfare services faced by budget constraints. These factors interact with current geopolitical circumstances, and the degree to which significant partners are committed to détente (e.g. through major East–West trade deals) to create possibilities for more or less conflict.

(b) Geopolitical factors

Events in the border areas of the superpowers are particularly important, with each superpower interpreting instability in its own sphere as reflecting the machinations of the other, and with its attempts at controlling instability being viewed solely as imperialist expansionism by the other. Where border conflicts become acute, and especially when control over important resources or access to strategic territories are implicatly involved, the prospects for East–West conflict increase. The role of the UN system and the Security Council in containing such conflict will depend upon the degree to which these international bodies are used in a principled, rather than opportunistic, fashion by the member nations.

(c) Military technology and doctrine

Although not unrelated to overall strategic planning, steps in military technology and shifts in the associated doctrine may themselves affect the level of political tension between the superpowers. The perception that one side may gain a strategic advantage that undermines deterrence based on mutually assured destruction may promote arms races and pressure for pre-emptive action, whether or not this perception is accurate. Proliferation of nuclear technology may be itself a source of tension, and the possibility of client states gaining access to weapons of mass destruction can be a reason for increased military preparedness.

(d) Internal politics

The military–industrial–bureaucratic complexes and the peace movements of both blocs can shape the political orientation of the superpowers and their junior partners. The power of the complexes may be determined by political struggle within the military establishment and government bureaucracies, whose results are conditioned by the power of different industrial and financial actors, and the strength and orientations of popular movements (especially those demanding increased services and consumption in place of unproductive expenditure). Peace movements, drawing on public concern about the threat of military conflict and also about the threats to democracy and civil liberties of the 'nuclear state', may establish new international channels and alternative forms of production and defence.

Table 3.4: *North-South relations: factors conditioning the types of world order sought by North and South*

(a) Internal political development

Nature of development strategies pursued by Northern countries will determine both their interests in world development and their ability to adapt to changing environments. Social movements and pressure groups around issues such as colonialism, disarmament, import controls, etc. could bring influence to bear on these strategies, and much heterogeneity may be anticipated — including attempts to imitate or avoid the experiences of other Northern countries. Political crises would mean that foreign policies are very variable and ineffective.

In the South, the displacement of dominant groups, presently tied to Northern interests through political and economic links, from positions of power, would tend to facilitate the formulation of more radical demands for global change. The pattern of national development in such instances — i.e. the social groups active in such changes, and the corresponding ideologies and programmes involved (e.g. socialist, populist, pro-East, etc.) — is relevant both to questions of foreign policy posture and to attempts to cooperate among Southern countries.

(b) Economic Factors

With economic restructuring of the North, the development of new technology may significantly affect interests in world development: resource-saving, labour-saving, resource-substituting technologies may render Southern countries less attractive as sites of production (if not as potential markets). Continuing recession might lead to more isolationist postures, though it is likely to enhance aggressive search for market outlets.

Fragmentation of Southern interests would suggest a divergence of attitudes to the world economy or a general tendency to acquiesce to Northern demands. To this extent postures proposing minimal changes would be likely to be chosen in order to muster maximal support, but in these circumstances it is quite conceivable that smaller groups of Third World countries — especially if their resource endowment or geopolitical significance gave them relative room for independent action — would form blocs either pressing for more significant change or engaging in more self-reliant strategies. Greater Southern unity would tend to strengthen attitudes to change, but excessive eagerness to gain recruits might produce a 'lowest common denominator' of weakness.

(c) North–South negotiations

Inevitably the types of demand raised by Southern countries, and the way in which they are raised (by whom, how abruptly, with what sanctions, etc.) influence Northern postures. Depending on conjunctoral factors the response may be retaliatory or conciliatory, designed to get rapid results or to stall for time, thus it might range from outright rejection of Third World demands to agreement to negotiate on main issues or to an offer of minor concessions in order to delay or forestall decisions. Intransigence might lead to stronger Third World demands and sanctions, depending on the power with which it is backed and the ability to restrain unity in Third World countries. Likewise compromise might lead to an escalation of demands in certain circumstances.

Other futures are possible, beyond those normatively defined by the three worldviews. Global disunity might mean a drift into deeper protectionism and a considerable extension of the world economic crisis. East–West tension might result in a spiral of conflict toward Armageddon. Other types of large-scale military confrontation are also plausible, especially along South–South and North–South axes, and these could quite easily inflame East–West relations.

The actual course of events will almost certainly fall between the alternatives outlined above. The point of scenario analysis is not to detail an infinity of variations or combinations of major themes. More

Figure 3.1: *Scenario for a liberal economic order*

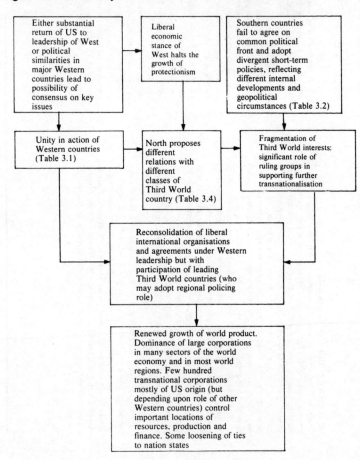

useful is to provide a view of key tendencies, the factors operating on them, and the logic of their development. The next section outlines a limited number of global scenarios, deriving from the discussion above. These scenarios are not predictions, but guides to the future, suggesting issues about which decisions have to be made if we are to have any influence over the one thing that is certain in the future — change.

A SET OF GLOBAL SCENARIOS

Figures 3.1 – 3.3 display how the three normative scenarios might be realised. The flow diagrams outlines future histories to which greater detail may be added. Let us take, for example, the route to the liberal ' economic order as depicted in Figure 3.1. How might this come about?

Figure 3.2: *Scenario for a reformed economic order*

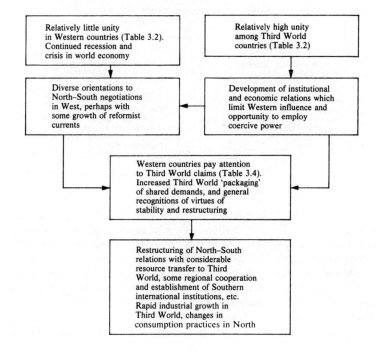

Figure 3.3: *Scenario for collective self-reliance*

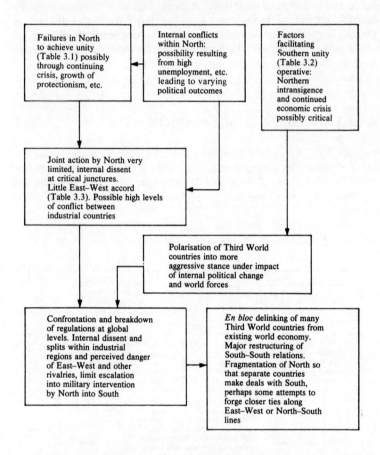

This scenario might involve something like the following sequence of events: Emerging from a round of trade summit negotiations in the near future, the main OECD countries agree to establish a system of international consultations, whereby the consequences of the economic policies of each can be debated and assessed, with the USA probably taking a firm lead in the discussions. From this emerges a system of closer cooperation in policy-making, whereby more co-ordinated actions can be taken to overcome the problems of stagnation and inflation in the world. Perhaps in exchange for assurances on help with readjustment, these countries would agree upon a common policy for closer economic integration and freer world trade which could be presented to the Third World from an unshakeable stance. This would require a firm basis for the policies of the western economies, ideally derived from large electoral majorities for conservative political parties.

The rich countries would have something to offer the Third World in exchange for their accepting the economic ground rules they are proposing, and refraining from disrupting raw material supplies and political order. There would be a phased (but not long-drawn-out) opening of the industrial countries' markets to goods (especially manufactured and processed goods) from the Third World. International organisations (such as the International Development Agency of the World Bank) would be given more funds to invest in developing countries. Rather than buying time by accommodating to the self-interested claims of Third World élites, the rich countries would seek to offer the world economy only sound guidelines and leads.

The world economic system which would be established through this scenario would, according to the conservatives, facilitate stable economic growth and reduce international tensions. Rather than depending on a few strong economies to pull along the remainder, the different industrial countries would rely on close communication and cooperation. As Third World countries followed this progress, their economies would necessarily become more outward looking. Manufacturing would grow fairly rapidly in these countries under the initiative of transnational corporations. (This might be further encouraged by permitting limited protection of 'infant' industries.) Investment would be increasingly free, with the funds provided by international agencies hopefully coming under a degree of Third World management through processes of divestment. These changes would call for more global consultations and cooperation, but this should not be allowed to develop into a regime of meddlesome international organisations. The scenario is viewed by its proponents as one of high technological progress, with stable energy and materials supplies and increasing trade and with gradual, rather than disruptive, changes in the patterns of production.

This account is formed very much within the logic of the conservative worldview. In order to draw a sharp distinction we will contrast this account with the radical scenario (see Figure 3.3). This scenario might involve a sequence of events somewhat like the following: After some years of further failure of North–South and UNCTAD dialogues to bring about concrete results, the strength of opposition to the established world order grows stronger. OPEC countries are frustrated by the dwindling of their revenues. Third World industrialists and military rulers increasingly resent the slow transfer of manufacturing industries to these countries, and significant numbers of the masses of people in hardship and near-famine conditions in these regions are radicalised. Many revolutionary struggles and popular movements against Western influence are strengthened. Cartels of raw materials producers take action to change the terms of trade (perhaps backed by OPEC price rises and selective sanctions). Demands for rapid agreement on changing the international order are met with compromises from the North, which is too divided internally either to coherently oppose or to acquiesce to such demands. With the half-hearted support of some industrial countries, much of the Third World effectively withdraws from some world organisations, and transforms others. After a period of uncertainty, new organisational structures may be negotiated which give significantly more weight to Third World interests.

Multinational corporations and other agents of international transfers would be subject to stricter regulation than in the past. Within the South, organisations would be established to enhance cooperative development: for example, in the integration of industries, regional resource transfers, divisions of labour, and research and development. There could be new world organisations, such as an international agency for technology and development, to focus development efforts on the problems of the Third World, and to act as an early warning system for technological innovations in the North. Terms of trade, aid and technical assistance would be improved for the Third World, which would experience faster, and more equitably distributed growth.

Various factors may render this scenario unstable, including possible divergence in short-term political interests, and rates of growth, of Third World countries. Also possible is concerted retaliation on the part of the North, which would otherwise have to undergo considerable economic readjustments, and perhaps very low growth rates for an extended period. As the circumstances of the Third World improved, the world might be more ready to accept fuller economic integration. In large part, this would depend upon political developments within the North.

Clearly, these two scenarios are not only mutually exclusive histories of the future, but are also based on different assumptions about how the world works. While we are not neutral among worldviews, each does need to be taken seriously. It is important to see what views of the future they project, and whether what each regards as the others crucial weaknesses, are liabilities or actually strengths.

Rather than enlarge upon the reformists' desirable future at this point (the account above should make it easy to envisage how a third scenario might be constructed) let us briefly consider some further alternatives. Each of the three scenarios depicts the future that might be attained were the normative vision of a worldview realised. More likely, however, is that the conflict among proponents of each worldview would undermine some of the conditions for realisation of each scenario, resulting in a stalemate, or a compromise, between strategies.

Two forms of stalemate can illustrate something of the range of compromise and conflict between the scenarios outlined above. One stalemate would be a *continuing global crisis*: a failure to marshal any scheme of economic regeneration and to extricate most national economies from economic instability and stagnation. Such a future might result from disunity in all world regions, with no escalation of superpower conflicts. It could evolve in many ways, one being the return of a more protectionist world divided into 'zones of influence', each dominated by their own local powers. A second form of stalemate between preferred strategies, is that of *global conflict*. This would involve the escalation and spillover of economic contradiction into political and military confrontation. Unhappily, a fairly high measure of military conflict is likely over the next decade in almost every scenario. But here we envisage a spiral of conflict based on a complete breakdown of East–West relations, perhaps triggered off by North–South conflict.

We need not enlarge further on these stalemate scenarios at this point. They round up our overview of possible global scenarios, which, we should reiterate, are presented not as predictions, but as outlines of alternative structural tendencies. The scenarios provide a basis for thinking about what combinations of events may arise, probably drawing upon more than one of these sets of tendencies. We now turn to thinking about scenarios for nations in the world system.

4 National Strategies and Scenarios

The three worldviews may present broad prescriptions for global change, but what of the strategies to be pursued by individual countries? Differences between countries are, of course, extremely important; the potential influence that countries can exert on the world system varies dramatically. The superpowers are capable of introducing huge changes in the world economy or poltical system, while many poor countries have a considerable task to even get their views registered. The prescriptions made from a single worldview for the same country are liable to vary with global circumstances. Figure 4.1 provides a simple framework for thinking about the relationships between international and domestic affairs. The exact nature of these relationships is liable to be seen quite differently from the three worldviews.

COUNTRIES IN THE WORLD ECONOMY

Because it is not possible to present analyses for every country in the world, it is necessary to categorise countries into a small number of groups to facilitate the analysis. What groups of country should we be concerned with identifying? Already the East–West–South trichotomy has proved too crude to make all the points required in discussing global strategies. We have sometimes identified specific countries. and sometimes talked about such subdivisions as newly-industrialising countries (NICs). The actions and interests of individual countries have considerable relevance to the concrete realities of power in the world, but can lead us into a rather short-term and superficial appraisal of events. In order to discuss longer-term tendencies it is helpful to shift our focus away from individual countries, and to consider groups of countries that are assigned distinctive locations in the world system. We hope thus to be able to

68

Figure 4.1: *The interdependence of international and domestic change*

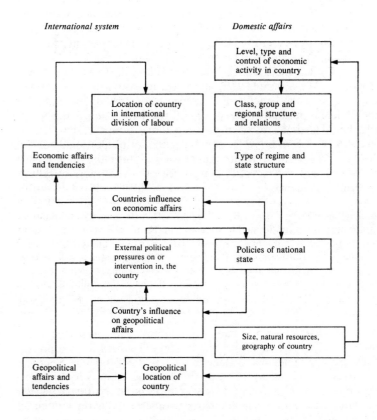

identify. and to model. major underlying processes in the international economy.

A disaggregation of world regions into different types. and also of national societies into different social groups. is implied. too. by the mutual critique of the worldviews' global strategies. Among the most significant criticisms of each strategy are those which point to the differential costs and benefits the strategy would imply for different types of country. Likewise. within countries, social groups receive differential outcomes: another reason for grouping countries into a few broad types is that the social structures within broad classifications like North and South are too diverse for us to gain much of a sense of the main social groups — and thus actors — within countries.

To talk of the South, for example, may obscure the existence of groups *within* Third World economies with varied and sometimes opposing orientations to the world economy; and differences *between* Southern countries, related to the patterns of development promoted by existing linkages, may be highly significant.

Who would benefit from a programme along the lines of the NIEO proposals? Weintraub (1979) attempted to assess the international distribution of the benefits of expanded trade, investment and more liberal debt terms and barriers to Third World products in the North. Dividing Third World countries into three groups based on their levels of *per capita* GNP, he concluded that disproportional benefits would accrue to the countries that already export more, have greater debts, and (with the exception of resource-rich countries) attract more foreign investment:

the key beneficiaries would be the wealthier |More than $500 annual *per capita* income| countries and to some extent the middle income |$200–500 *per capita* income| would be minor beneficiaries. For the poorest countries in the foreseeable future, levels of official development assistance will continue to be more significant than the integrated commodity programme, debt moratoria, the various systems of trade preference, the multilateral trade negotiations, new initiatives on technology transfers, promotion of private direct foreign investment, or better access to world money markets... (p. 255).

We would like to submit a range of different strategies to this sort of analysis, using the scenarios and model developed for the present study. For example, in a strategy of collective self-reliance, Southern countries would also obviously be starting from very different positions – with different levels of access to Northern money (resource-rich countries as compared to others), industrial structures, and labour supplies. Historically, regional integration of markets in the Third World (admittedly without delinking, so that benefits have often accrued to Northern transnational corporations), has reinforced existing differences in growth rates (Cable, 1980). It is easy to conjure up a scenario whereby the oil-exporting countries coerce the poorer agricultural-exporting countries to supply them with cheap food — holding out the alternative of purchasing, say, US grain — thus perpetuating the low wages and stagnation of large areas of the South and reproducing 'centre–periphery' relationships within the South–South context.

A more detailed classification of countries is needed in order to assess such differential outcomes of scenarios. In Chapter 3, we were already implicitly doing this. where. as well as the East–West–South distinction. we made mention of NICs. rich oil-exporting countries. and different levels of technological and industrial performance within the West. These categories may not capture all of the important points of difference between countries in the world today — far from it — but they are relevant to other locations of nations in the world system.

We distinguish between three types of 'core' or 'industrialised' Northern countries: these are the relatively high and the rather poorer performers in terms of industrial technology, and the centrally-planned economies who stand somewhat outside the capitalist world economy. while not constituting a parallel socialist world system. Likewise. we identify three types of 'peripheral' or 'developing' Southern countries: these are the 'semi-peripheral' NICs. resource-rich nations. and the less favoured exporters of primary produce. Of course. we have already needed to draw finer distinctions on occasion. For instance. in many respects. the superpowers are special cases; likewise it is obvious that the circumstances of 'giant' countries (such as India. and the People's Republic of China) face very different opportunities and constraints compared to the smaller nations which we assign to the same group.

We assign countries to the six groups as follows:

Group 1 comprises the six OECD countries with highest *per capita* industry-financed expenditure on research and development (Pavitt, 1980). which means that they are likely to be relatively innovative in technology. The USA is a member of this group: it assumes a position of technological and military leadership for other Western industrial countries (although the former is challenged by Japan, in particular, and the productivity growth of the United States is inferior to that of the other members of the group of countries).

Group 2 includes the remaining non-centrally-planned industrialised nations (according to the World Bank classification). It contains both a number of early industrialising countries. such as the UK. and other OECD countries. whose general economic performance is not outstanding among Western countries.

Group 3 consists of industrially advanced centrally-planned economies. The USSR occupies a leading political and economic position in this group. In addition to the Warsaw Pact countries. we include the 'market socialist' Yugoslavia because. though it is a 'special status OECD country'. its political economy is very distinct from that of Western countries. (Romania and Albania are not included in this

Table 4.1: *Classification of countries*

Group	Countries included	Notes	Group	Countries included	Notes
1.	Federal Republic of Germany, Japan Netherlands, Sweden. Switzerland, USA.	USA unusual because of its size. political dominance. and poor civilian technological performance compared to others. Japan unusual because of non-European culture, and relatively low *per capita* GNP and other attributes of 'early late industrialisation'.	6a	Bangladesh, Benin, Bhutan. Bolivia, Barman, Burundi, Cameroon, Central African Empire, Chad. Congo, Dominican Republic. El Salador, Ethiopia. Ghana, Guatemala, Guinea. Haiti, Honduras, Ivory Coast, Jamaica, Jordan. Kenya. Lebanon. Lesotho. Liberia. Madagascar, Malawi. Malaysia. Mali, Mauritania. Morocco, Nepal. Nicaragua. Niger. Pakistan, Panama. Papua New Guinea. Paraguay. Rwanda. Senegal. Sierra Leone, Somalia, Sri Lanka, Sudan. Syria. Tanzania, Thailand. Togo, Trinidad and Tobago, Uganda, Upper Volta. Yemen Arab Republic, Zaire. Zambia. Zimbabwe. Ecuador. Costa Rica. Egypt, India. Indonesia. Peru. Uruguay	The last seven countries have considerable industrial development. The group as a whole has a large range of *per capita* GNP and population.
2.	Australia, Austria, Belgium, Canada. Denmark, Finland, France, Ireland, Italy, New Zealand, Norway, South Africa, United Kingdom	South Africa unusual because of apartheid-based dual economy with pockets of severe poverty. and the major role of resource exports (especially gold). USSR unusual because of size and political			
3.	Bulgaria, Czechoslovakia, German Democratic Republic, Poland, USSR. Yugoslavia	dominance. Yugoslavia has a distinctive political process and planning system.			

72

Table 4.1 *Continued*

Group	Countries included	Notes	Group	Countries included	Notes
4.	Argentina, Brazil, Chile, Columbia, Greece, Hong Kong, Israel, Republic of Korea, Mexico, Portugal, Puerto Rico, Singapore, Spain, Taiwan, Turkey	Argentina, Chile and Colombia appear to have suffered severe setback to their industrialisation. Asian economies have better weathered the crisis.	6b	Afghanistan, Albania, Angola, Cambodia, China People's Republic, Cuba, Korean Democratic Republic, Lao PDR, Mongolia, Mozambique, Romania, Vietnam, Yemen PDR.	With the exception of Romania, these countries are exceptionally poor. China is a giant in population, resource and industrial terms.
5.	Algeria, Iran, Iraq, Kuwait, Libya, Nigeria, Saudi Arabia, Venezuela	Group ranges from the very rich near-microstate of Kuwait to the large and relatively poor Nigeria. Countries share high oil exports (as dominant export).			

73

group: they are treated as Third World countries because, on account of their low levels of industrialisation, both had less than a third of their labourforce in industry, and their *per capita* GNP was well below $2000 in 1977, according to World Bank data.)

Group 4 consists of NICs, who might be termed the "upwardly mobile semi-periphery" in world system theories. Ten countries are classified as 'semi-industrialised' by recent World Bank *Development Reports*, and these form a satisfactory set for our purposes. They include a number of Southern European countries as well as the Latin American and South–East Asian NICs. We do not include here the giants India and China, whose industrial capacity is large, and who have developed rapidly, but are still dependent on the large rural sector, and have relatively low *per capita* incomes.

Group 5 includes OPEC member countries (other than those belonging to previous groups). Oil revenues have placed these nations in a unique position in the contemporary world economy: some have *per capita* incomes equal to these of the richest countries, and others are outstandingly wealthy in their regions.

Group 6 contains all Third World societies excluded from previous groups, and spans the World Bank's middle and low income types. It includes both market and centrally planned economies, which may be treated as significant variants (Groups 6a and 6b).

Table 4.1 outlines the assignment of countries to the different groups, and presents a few brief notes concerning countries that are a typical of their groups. We shall not attempt to delineate country-specific scenarios in this book, but it is obviously necessary to proceed with particular caution in applying scenario and model results for a particular group of countries to these exceptional cases.

Table 4.2 and 4.3 contrast the country groups in terms of some social and economic indicators. Table 4.2 shows clearly that the industrialised countries rate above the less industrialised countries on most of these conventional indicators of development, such as life-expectancy or levels of literacy, although the standard deviations for each measure show that there is considerable overlap between groups. Table 4.3 shows indicators of public and military expenditures: while the proportion of GDP allocated to medical services also follow this trend, there is no clear trend for public expenditure, and military expenditures are greatest in the developing economies.

Our next task is to outline the circumstances and relations of these six different types of countries. On this basis we shall be able to move toward identifying strategies and scenarios for countries of each type.

Table 4.2 *Indicators of development by economic group*

Economy group	1	2	3	4	5	6a	6b
Life-expectancy at birth (1977)	74.0 (1.4)	72.1 (3.2)	70.9 (1.3)	67.5 (5.0)	56.1 (7.7)	5.2 (8.8)	56.2 (11.9)
Urban population % Total (1975)	74.2 (10.5)	68.7 (15.2)	56.6 (11.3)	62.9 (21.3)	56.2 (21.3)	28.2 (18.0)	29.5 (18.3)
Adult literacy % Total (1975)	99.0 (0.0)	97.2 (6.4)	95.0 (6.7)	80.7 (9.8)	54.4 (17.9)	43.4 (26.5)	64.0 (41.2)

Note: Data are unweighted averages based on United Nations sources for closest year available. Figures in parentheses are standard deviations for our samples.

Table 4.3: *Indicators of public expenditure by economic group*

Economy group	1	2	3	4	5	6a	6b
Public consumption % GDP (1977)	18.0 (6.0)	18.3 (2.7)	9.3 (2.3)	14.0 (7.9)	17.8 (2.7)	15.8 (7.1)	20.3 (6.0)
Defence spending % GDP (1977)	3.4 (1.8)	2.9 (1.4)	5.2 (3.6)	6.1 (8.7)	11.0 (9.6)	5.8 (6.6)	7.8 (4.0)
Inhabitants (1000s) per doctor (1970)	7.2 (1.1)	8.4 (3.7)	6.0 (1.9)	13.8 (7.0)	69.0 (67)	207 (228)	112 (67)

Note: Data are unweighted averages based on United Nations sources for closest year available. Figures in parentheses are standard deviations for our samples.

CHARACTERISTIC OF WORLD REGIONS

This section mainly concerns two types of characteristics of countries in the different groups. The first characteristics are economic ones, in particular, certain features relevant to the model which we use later in this book. The second characteristics concern major actors in the different countries, which are relevant to their interests and programmes which can give rise to different scenarios.

The first set of issues receives most attention in official outlines of development prospects. These reports confront us with familiar statistics on economic and demographic growth, and the structure of production and trade. More recent reports may give some opportunity to infer the constellation of group interests and power relations in a country by providing data on the distribution of income or consumption levels, and it is possible to reach conclusions about the implications of accounts of 'resolute' or 'shifting' economic policies for the balance of power. The economic features such reports illuminate are very important determinants of future prospects; however, we need to supplement them with more explicitly political and sociological considerations.

For scenario analysis, as outlined in Chapter 2, we need to assess the experience of different groups and group interests. Any one country, of course, contains a vast diversity of distinct social groups, and social cleavages may exist along age, ethnic, religious, regional, urban–rural, and many other dimensions. All of these may structure different images of the future. How then are we to identify a manageable set of groups?

One fundamental set of social cleavages is structured around class differences. Class relations have been studied in detail: the three worldviews each pay attention to the definition of social class, and the role of class conflict and cooperation in achieving change. Further, the worldviews are typically advanced by representatives of different class interests, even though they may claim to represent global or national interests.

Social classes may be identified in terms of the different locations of people as economic agents to the means of production – this is closest to the radical definition of class. For example, some people subsist by their own production with their own land and equipment; some work for wages on farms, in factories or faculties owned by others; some are involved in the distribution or management of the social product rather than in its original production, and again their positions may be those of owners or workers. Thus, people can be assigned to different classes: peasants, wage-workers, owners, and the like. Different types of country have their own characteristic mixtures of relations of production, and thus distinctive class structures.

In concrete circumstances, class formation and class relations are crucially affected by the state. Its role is to use political means to facilitate the functioning of the economy, to manage and reproduce at least some of the relations of production of the society, and this necessarily affects the status and orientations of different classes. The state's actions are typically structured by the interests of a coalition of the dominant social groups in a society. Across different groups of country, states may be more or less open to demands from different agents: in some societies there is considerable freedom to organise pressure groups and make use of pluralist institutions, while in others, state power takes authoritarian, military or paternalistic forms.

Next we describe a preliminary characterisation of the different world regions in terms both of conventional economic data and of class analysis. Our account here draws on a wide range of sources, including annual compilations of economic data like the World Bank's *Development Reports*, and non-official publications like Pluto/Maspero's *World View* (1981). Useful analyses from world system perspectives includes Amin's *Unequal Development* (1979), and Chirot's *Social Change in the Twentieth Century* (1977).

THE NORTH

We first outline some common features of group 1 and 2 countries before focusing on the factors that lead us to differentiate between these groups. These are mature capitalist societies where most pre-capitalist economic relations of production have been displaced or, at least, transformed by industrialisation and the growth of the 'service economy'. These societies have not fulfilled the classical Marxist prediction of class polarisation, which expected that they would come to be composed of only two classes — industrial workers and the owners of large businesses. Instead, they possess substantial petit bourgeois groups, such as small shopkeepers and tradespeople who own their own shops, equipment or professional skills, but are basically playing service roles in larger circuits of commodity production and exchange. The numerically dominant classes are the traditional working classes (in industry and, increasingly, in services) and the middle classes of workers with established career structures, technical or bureaucratic expertise and relatively high control over their immediate work. The middle-class group is possessed of skills, intellectual resources, and the tools of the trade, and is more affluent than the working class, but has little control over the major productive resources of the society. Both groups involve many women workers, who tend to fill lower occupational positions due to the double burden of employment and domestic commitments as well as sexual discrimination.

The means of production and distribution are increasingly controlled by managers and owners of large corporations, where there is a complicated pattern of interlocking directorships and financial stakes within and between financial, productive and distributive concerns. A very small proportion of the population actually owns the major assets of these corporations, although there is a considerable number of small business owners. There are effectively four major economic structures operating in parallel: the corporate economy, the competitive business sector, small firms and the self-employed, and the productive, administrative and welfare agencies of the state.

The states in these countries have considerable control over their domestic and foreign affairs, which means that they are less likely to exercise force to secure their objectives. These states have, typically, achieved political and economic consensus through large-scale welfare state activities. These are now increasingly subject to financial restrictions, despite the trend towards stronger working-class and middle-class organisation. While the dominant coalition has tended to dictate the shape of government policies, it has been bound to take into account the pressures from different class interests, particularly in domestic policy. The dominant coalition in the state has typically been representative of manufacturing and commercial interests (with financial groups apparently gaining more ascendancy with the rise of monetarism). This coalition typically supports domestic industry, and tends to gear production for home markets, and to use state power to gain access to foreign markets. The state has historically moved in to support declining industrial sectors, and with the current wave of technological change has played an important role seeking to promote technical innovation and rationalisation.

Group 1 countries, the most powerful and technologically advanced industrial market societies, have a number of common features. They are important in the OECD organisation of Western interests, and thus play a major role in structuring North–South relations. They are dependent on oil and raw materials imports. Growing rivalry exists between them, especially around the penetration of Japanese imports into the domestic US and EEC markets. This rivalry may be settled by negotiated settlements (e.g. joint ventures, specialisation) or flare up into damaging trade wars.

The internal politics of these countries has been relatively stable, following extensive restructuring after the second world war. Agriculture has largely been modernised and much of industry is equipped with advanced capital stock, although peripheral industrial sectors service the core industries to a considerable extent. Internal markets are typically large, with convenient access to foreign markets. Relatively high wages provide consumer markets (supporting product innovation) and motivate process innovation (Japan, however, combines high innovation levels with relatively low living standards). A large proportion of exports are manufactured goods, balance of payments are relatively favourable (although often negative), and there are considerable international reserves and levels of overseas investment.

Group 2 countries share many characteristics with Group 1. Both are highly industrialised and urbanised, with relatively high income levels. Although their consumption patterns are linked to those of Group 1, Group 2 economies are less competitive in world and domestic markets for most technologically sophisticated products. This means that these countries are performing relatively poorly in the world economy, as reflected in balance of payments and other problems. (In turn, this makes it harder for this group to obtain secure supplies of raw materials.)

The economic strains may be taking their toll on traditionally stable parliamentary systems. Many observers had already seen these as slowly evolving into a more corporatist direction, with complex management tasks increasingly being jointly negotiated by civil servants and representatives of affected group interests. But political polarisation along class and regional lines is increasing in many of these countries, as economic restructuring conflicts with social goals, and it is possible that coporatist tendencies will be overtaken by dualistic ones.

Group 3 countries also show signs of political strain, with their relatively monolithic political institutions finding it difficult to adapt to a range of new problems. The chief economic problems are posed by the need to shift away from growth based on extensive development. Previously, the expansion of the labourforce and the exploitation of new land areas enabled steady improvements in output levels and quality. Although growth has tended to be higher than in Group 2 countries, agricultural development has had many setbacks, use of capital stock is relatively inefficient, and consumption levels have been relatively austere. (This has created problems in motivating workers, and in Eastern Europe people have contrasted their situation unfavourably with the self-portraits of the Western media.) Considerable raw materials reverves in the USSR (which has to date provided other CMEA countries with energy at concessionary terms), require imported technology to develop fully: the high growth they have permitted in the past is unlikely to be sustained in the immediate

future. Primary products and resource-based manufactures are traded
in exchange for high technology. These manufactured exports often
compete with those of Group 4 countries, and high external debts may
increasingly limit the ability to subsidise exports. Trade and aid links
with the Third World mainly involve a restricted set of countries, and
are often based on political sympathy.

The political economy of Group 3 countries differs sharply from
that of Group 1 and 2 countries. In Group 3 societies, through internal
revolution or the division of Europe after the second world war, state
power was seized by previously subordinated social groups. These
groups were intent on achieving national development outside, rather
than on the periphery or semi-periphery, of the capitalist world
economy. Distinctive social structures have been established, despite
the considerable similarities in occupational composition shared with
other industrial states. Group 3 countries lack classes that subsist on
the rent of capital or property; and private ownership of businesses is
mainly restricted to small operations in agriculture, and consumer
goods manufacturing and distribution. This class structure is reflected
in a relatively egalitarian distribution of wealth, although fairly large
income differentials and a host of privileges are associated with
occupational status. With the state controlling the means of production, but relatively insulated from popular demands, bureaucratic
occupational groups form a large and significant élite.

While the bureaucracy forms a more homogeneous group then the
élites of Group 1 and 2 countries, working-class interests tend to be
fragmented and individualised. Independent trade union organisation
is largely prohibited because state control of production means that
economic changes are liable to be rapidly politicised. Political
organisation is generally dominated by the state, and thus the
dominant coalition is one of bureaucrats, upper management and
professionals, and military leadership. The main social classes, then,
are a large industrial and agricultural working class, with less
organisation than in Groups 1 and 2; a middle class which is relatively
closely tied to the state apparatus, and the dominant coalition. The
state has tended to limit consumption levels, focusing on the
development of heavy industry in a drive to catch up with the West.
However, the conservatism of the bureaucracy has made technological innovation difficult.

THE SOUTH

Group 4 countries have complex class structures, partly reflecting their different geopolitical locations and cultural heritages. The Southern European economies have a great deal in common with Group 2 countries, but are at lower levels of industrial development. The East Asian 'boom' economies and the richer Latin American countries have been 'modernised'. New centres of social power have been rapidly established around dynamic capitalist industries in societies which had recently been largely organised around the interests of 'feudal' landowners. Further differences are grounded in the (typically Asian) export-oriented versus (typically Latin American) import-substitution strategies for industrialisation which have been pursued over much of the post-war boom.

While the dominant coalitions reflect these historical processes and choices, the state in both types of NIC plays an important active role in the economy. It is typically interventionist in organising productive industry, and limiting wage levels, often by quite repressive means. There are major income inequalities, with large middle classes coexisting alongside masses of working poor and marginalised groups. The southern European NICs too, have had considerable recent experience of military–authoritarian rule, but are generally characterised by rather better living standards and welfare services.

Political stability is very variable across these countries, and the dominant coalition, typically, represents a mixture of divergent interests which helps account for the political volatility which several display. As well as state officials (who are liable to develop their own interests), there are industrialists and financiers allied to transnational capital, and those controlling national capital. The state is thus involved in mediating between, and making choices among, diverse demands: its military arms or bureaucracy may face important choices between relatively left-and right-wing development strategies. With less global and national power than the states of Groups 1 – 3, the governments of Group 4 countries are more likely to have recourse to the use of force in domestic affairs.

Apart from the dominant interests, there are fairly substantial and economically privileged middle classes, and a large working class (often including a large marginalised sector of unemployed and 'informal sector' workers). There may be a large number of peasants, but these are a declining percentage of the population. Working-class organisation is advanced in some of these countries, but is often subordinated to the state through official trade union structures.

Southern European economies have a great deal in common with pluralistic political parties, and Latin American countries also tend to share some measure of working-class political organisation.

We identify these Group 4 NICs by their rapid growth of industrial production and manufactured exports (which pose a challenge to the products of other groups). They generally possess quite advanced capital goods industry and infrastructure, displaying a capacity to absorb and develop new technology which, together with depressed wages, gives them a competitive position in world markets. Several of these countries possess a strong base of natural resources, and most have developed considerable technical expertise. This has enabled them to cope well with the structural adjustments subsequent to the oil crisis — if not with the burden of extensive debts. (A few of these countries have accumulated the larger share of Third World debt, and the considerable strain of simply servicing these debts has led to quite dramatic revisions of their development plans.)

Group 5 countries occupy a unique location in the world system because of their oil wealth. In the case of the Near East and Gulf countries, the major inflows of capital have strengthened, but also transformed, traditional élites. While maintaining state power, these groups have typically pursued 'state-capitalist' development strategies. The dominant coalition has been very limited, with ownership concentrated in the hands of small and often nepotistic groups. States have sought to ward off popular discontent through welfare and other expenditures, while rigidly circumscribing political activity. Yet urbanisation, and the growth of merchant classes associated with affluence and urbanism, have eroded the traditional authority structures connecting the rural poor to the central leadership. Islamic revivalism has formed a basis for social criticism (rather as the Catholic Church provides in some Group 3 and 4 countries an institutional framework, and set of values, in which political alternatives can crystallize). The oil boom has recently lost much of its momentum with the world recession and energy conservation policies.

The Near Eastern oil states exist in a particularly charged regional situation. The question of Israel and the Palestinians continues to dominate Near Eastern politics, imposing its own cross-cutting set of alliances and oppositions on the Arab states. Some members of Group 5 are perhaps more similar to other groups in their social and political structures. For example, Venezuela is a relatively industrialised Latin American economy, in many ways resembling that region's Group 4 countries; Nigeria is in many respects a resource-rich African Group 6 country.

The economic performance of Group 5 countries has been uneven in recent years, with a glut in the oil market reducing their gains from oil exports. It has proved more difficult than anticipated to industrialise on the basis of oil wealth. Agriculture and industry are relatively weak, although there has been considerable development of infrastructure. But these economies remain reliant upon foreign expertise and, in some cases, on migrant workers. Their balance of payments and foreign holdings are usually strong, as is reflected in the high *per capita* incomes. But the latter statistics disguise large inequalities that may fuel political instability. Prestige investments and improved public services, together with the diversification of industry, are part of an effort to head this off.

Group 6 countries are tremendously diverse, united more by poverty than positive features: they lack wealth, are relatively poorly incorporated into the world market and have weak states. But the degree of material poverty still varies tremendously within this group, which can be classified into various subgroups: resource exporters and agriculture exporters, large and small countries, etc. We distinguish here, particularly, between the market and centrally-planned political economies, whom we term Groups 6a and 6b respectively. (Several cases are difficult to assign with much confidence, and political upheavals may move countries from one path to another, however. Because data are especially weak for these countries, in our economic model we shall treat them as a single group.)

We consider first the market-based Group 6a economies: most of Africa, and the poorer Latin American and Asian countries. The dominant coalition in these countries typically represents economic interests oriented to the world market. They may be rooted in landownership, primary production or (less often, but increasingly) in manufacture for exports. A few countries are dominated by pre-capitalist interests, exerting their power through 'feudal' social arrangements. Almost all of these countries possess large peasant classes involved in traditional systems of production, though these are coming under the grip of market demands.

Agriculture produces only a small investable surplus which is often appropriated by urban groups, leaving the basic needs of peasants and landless workers unsatisfied or barely met. This leads to population growth and rural-urban migration. The level of development of national capital is low and mostly organised toward luxury products, with little orientation of industry to support agricultural development. Transnational capital has entered countries to different extents, sometimes to exploit primary production on a large scale (plantations, mining, etc.), sometimes in localised industrial projects. Exports have been plagued by declining terms of trade and price instability.

There are small middle-class groups, to a large extent composed of state employees. The urban working classes are also a relatively small proportion of the population (and accompanied by many marginal workers, 'informal sector' workers, and unemployed people). There are large numbers of peasants, in a variety of landowning relations, and often living in acute rural poverty; there may also be rural plantation or ranch labourers. Whatever the dominant coalition, the rural power structure makes it difficult for rural workers to better their positions. There is growing unemployment, low wages, and restricted demand for basic goods. These features are shared by most Group 6a countries, although basic needs may be better met in the middle-income countries. The alliances in their dominant coalitions are various: there are 'neo-colonial' states based on traditional landowning power, or serving the interests of local agents of transnational capital. Other countries are pursuing more liberal or state–capitalist strategies, in which national capital plays a more dynamic role.

Both Groups 6a and 6b share many of the attributes of underdevelopment. Productivity and capital stock are low, the economy is often reliant upon agriculture or some raw material. The economic and political infrastructure is often weak, with considerable dualism between town and country. The rural areas typically feature subsistence production, minimal consumption among low-income groups, and a subsidisation of urban consumption and production by agriculture. Group 6a governments are often military, and population growth and urbanisation are typically rapid. Group 6b countries have made considerable efforts to reduce inequalities and increase consumption of the poor – with some success – and they are less influenced by transnational corporations in their export sectors.

Group 6b countries have typically sought to break from the development problems exprienced in Group 6a, by means of peasant and military-based revolutions, such as those in Angola, Vietnam and the People's Republic of China. While the length of experience with their new social orders is often very short, and while most of these countries have faced immense tasks of reconstruction after civil war and foreign intervention, a number of common features are apparent. The poorer countries remain largely peasant-based societies, although the collectivisation of agriculture increasingly gives peasants the characteristics of a rural working class. There has been considerable effort towards heavy industrialisation, giving rise to an industrial working class, and layers of skilled workers; some of these countries

(Romania, North Korea) resemble NICs in this respect, although their trade patterns are very different from that of Group 4. The state apparatus, in which military groups often play a leading role, has grown considerably, and is far more powerful than that of Group 6a. Its dominant coalition represents a fusion of several interests. While initially grounded in a popular party representing workers' movements, the state administrative and military groups have developed their own distinctive interests, which much of their action may be seeking to defend.

NATIONAL SCENARIOS

Having outlined some of the main features and potential lines of social cleavage of the different country groups, we can go on to contrast scenarios for each group. We begin by discussing Groups 1 and 2 together, since they share so many features.

Political debates within Group 1 and 2 countries present a wider range of strategies than could be accounted for by the diverging interests of capital and labour, though this basic principle of their class structure remains central to the viability of different strategies. Figure 4.1 outlines some of the forces that play a key role in establishing the balance of power between capital and labour. These show important cleavages within the dominant coalition on several issues: (i) state support (such as protection) for declining industrial sectors; (ii) state support for technological innovation in new industries (through, for example, military contracts and research and development provisions); and (iii) finance (the level of public expenditure and control of the money supply). Large and small businesses also have distinctive viewpoints. It is not just industrialists who are divided: parts of the labour movement, aligned with the defence of traditional industries, support import controls and reflation. Others, especially those employed in public services, are more concerned with supporting the welfare state apparatus. Middle-class groups seek greater access to control of bureaucracies and corporations alike. These complex patterns of interest could, in principle, lead to new coalitions which may gain control of state power.

We can identify four scenarious representing fulfilment or compromise of different groups' programmes (see also Miles, 1983):

(1) *Corporate society*: an economic restructuring aimed at restoring rapid growth, favouring large firms, stimulating growth through new technology and a reduction of 'unproductive' expenditure

Figure 4.2: *Scenarios for Group 1 and 2 countries*

	Corporate society	*State capitalism*	*Democratic-socialism*	*Decentralism*
Main proponents	Financial capital and sunrise industries, allies in the state (e.g. Treasuries and bank officials)	Declining industries, traditional working-class unions, labour movement bureaucracy.	Some public sector unions and workers, radicalised fractions of labour movement, some parts of local state.	New middle classes, and the social movements around them (peace, ecology, feminism) together with some minority groups.
Conditions of realisation of programme	Factors facilitating industrial and financial coalition active, those favouring working-class movements inactive. Political and cultural support for economic restructuring. Continued successful innovation of new technologies.	Failure of financial interests to mobilise resistance to state intervention, successful labour–management coalition in declining industrial sectors. Weakness of 'new left' currents in labour movements. (Probably, failure of solidarity among Northern states, leading to strengthening of protectionist tendencies.)	Weakening of dominant coalition in face of upsurge of labour movement activity, probably requiring classic 'revolutionary crisis' of legitimacy of social order; leading to struggles establishing considerable measure of worker and community control over means of production. Requires considerable international support.	Stand-off of labour and capital, and their traditional political parties, permits social movement and regionalist currents to play decisive role in politics. Steps tolerated or welcomed by traditional political groups, creating hard-to-reverse political recomposition

Figure 4.2 *Continued*

	Corporate society	*State capitalism*	*Democratic-socialism*	*Decentralism*
Production and trade	State uses contracts and subsidies to support new technology. Effort to diffuse most productive equipment in industry: displacement of labour, at least in medium term; demand for more technicians at same time as deskilling of many jobs. Liberal trade policies.	Corporatist intervention in major economic sectors, with union leaders providing discipline of workforce. High levels of public works, reduced working week, protectionist trade strategies: very patchy innovation.	Efforts to regulate production according to social needs, to reduce working week and humanise workplace. Boost to basic production and social infrastucture (e.g. using new communications technology to coordinate production), and attempt to develop technology capitalising on existing workforce skills.	Development of 'informal' and local production for direct use reducing proportion in formal employment. Resource-conserving technology, somewhat decreased trade due to 'self-reliant' orientation.
Distribution and consumption	State expenditure and taxation limited: increased inequality, especially between wage-earners and non-earners. Considerable mass consumption of cheap electronic goods, for privatised mass communications.	Relatively slow growth of private income and luxury consumption, high taxation, emphasis on state provision of some services and basic goods. Attempts to regulate price structures.	Emphasis on collective provision of services and new infrastructure. Regulation of prices and incomes, greater equality and higher consumption levels of poorer groups.	Efforts to reduce levels of wasteful consumption and to place welfare services under community management. Greater consumption of locally-produced, labour-intensive products.

87

Figure 4.2 *Continued*

	Corporate society	State capitalism	Democratic-socialism	Decentralism
Political structures	Growth of repressive apparatus to limit expressions of discontent. Limited control by public over major economic programmes. some corporatist decision-making introduced. Strong emphasis on individualism.	Some participatory structures of limited scope established. populist approach to cultural affairs and egalitarianism in welfare delivery services: but generally bureaucratic regulation of key economic and social choices.	Experimentation with systems of direct democracy and popular planning, often reliant upon organised social movements as active participants. Greater role of producers in determining overall allocation of resources of local groups in determining use to which put.	Community organisations and local authorities more important in national planning. Participative planning systems established. redirection of civil service under 'anti-expertise' orientation.
Major dilemmas	Difficulties in generating employment with new technology, leading to problems of unemployment and economic dualism. Resistance to declining living standard and repression. Campaigning on behalf of declining sectors, against militarism.	Resistance to centralism manifest in various ways — e.g. black markets, wildcat strikes. Sabotage by withdrawal of capital (capital strikes, exports) and, perhaps, mobilisation of paramilitary opposition and of discontented groups in armed forces.	Relating together local popular planning iniatives to requirements of complex interdependent economy. Overcoming resistance of displaced dominant groups without resource to repressive practices.	Articulation of high technology sector with small-scale production. while attempting to avoid substantial power imbalance in favour of, or dependence on, the former. Risk of stimulating dualism and segmentation which would establish different working conditions and privileges. Shifting public taste without creating resistance.

and consumption; requiring the passive compliance of, or tightening of legal controls on, working-class organisation.

(2) *State capitalism*: major intervention of the state into the affairs of productive capital, with appropriation of major institutions and attempts to control investment and product development in the light of market criteria *and* long-term national economic competitiveness.

(3) *Democratic-socialism*: nationalisation of large firms and the establishment of new channels for popular representation and control of political affairs and the local economy and workplace.

(4) *Decentralism*: major reorientation towards local production and the informal economy, together with partial nationalisation of private capital and attempt to limit state cntrol of everyday life.

Figure 4.2 contrasts salient features of these four scenarios. As was the case for our global scenarios, the future is likely in practice to involve some combination of these different programmes of development. The failure of one programme may lead to an attempt to instal others. (We have not set out to portray scenarios relating to the more vicious backlashes (e.g. Fascism) that might accompany such circumstances.)

While the scenarios are not all equally likely, and different programmes have different prospects of success in different countries, the scale of the current economic restructuring means that it would be unwise to rule any of these projects as outside the agenda. It is quite possible that the development strategies of various countries in Groups 1 and 2 will diverge, which affect both the viability of the different programmes, and the evolving global scenarios. 'Corporate society'-type scenarios favour Northern pressure for a liberalised world economy, while others could facilitate a new international order, or strategies of collective self-reliance.

Group 3 countries

In many respects, these countries are living in an impoverished version of the second of the Groups 1 and 2 scenarios (although it is less subject to market forces and more accurately described in this case as 'bureaucratic collectivism' than as 'state capitalism'). The programme currently espoused by the dominant coalition in these societies envisages a smooth transition from extensive to intensive growth, marked by the 'scientific–technological revolution' in production, decentralisation of economic decision-making, raised living standards, and poltical liberalisation. The likely alternative outcomes include: a growing economic crisis, or an increasing effort to use market mechanisms to overcome these problems; finally grassroots social movements may be established and play a role in the course of development. We identify the corresponding scenarios as:

(1) *Scientific–technological revolution*: the official vision of the future, marking a substantial step towards an affluent communist society.

(2) *Stagnation*: an unstable stand-off between movements for reform and centralisation, with declining economic performance and growing political alienation.

(3) *Reform*: a drive to decentralisation of economic decision-making and adoption of Western technologies and economic practises.

(4) *Democratisation*: growth of self-management and liberalisation currents, leading to the consolidation of centres of power more independent of the state apparatus.

Figure 4.3 contrasts key features of these four scenarios, (which are derived in part from the work of Nuti (1979)). Again the probability of the different scenarios is far from equal: different Eastern European countries vary considerably in the degree to which social movements can hope to establish alternative bases of social power. As for the fulfilment of a 'scientific-technological revolution', past performance suggests that it is not easy for the state to reintegrate individuals into social activity after they have been socialised to accept an atomised existence in the name of 'socialist construction'. Perhaps a likely outcome would be a continuation of the historic fluctuations between centralism and reform in the East – but global economic restructuring may render such fluctuations unstable.

Group 4 countries
These semi-peripheral countries feature a range of histories and social formations, although, commonly, the state plays an important role in mediating between national and transnational capital. This can lead to a number of quite different structures of the dominant coalition. Clearly, the power of the different groups involved will depend upon past strategies, but there is little reason to expect a mere continuation of past models.

Here we identify four scenarios (drawing on a set of scenarios for Brazil outlined by Singer (1979), and a more general discussion of Third World class alliances by Petras (1975)), which reflect the dominance of different coalitions:

Figure 4.3: *Scenarios for Group 3 countries*

	Scientific revolution	Stagnation	Reform	Democratisation
Main proponents	Higher state officials and party members.	Result of self-interested and conservative behaviour of higher state officials, especially military.	Junior officials, middle management, intellectuals.	Working class groups, intellectuals.
Conditions of realisation of programme	Favourable international environment (detente, world inflation under control); improved generation and absorption of technical innovations, high level of social accord.	Continuing problems with industry and agriculture, possibly reflecting military dimension of resources; containment of worker demands for high consumption.	Failures of attempts to rationalise or rejuvenate central planning (e.g. with use of computer techniques); growth of regional sentiment.	Social movements establish independent bases of organisation, problems in Soviet bloc such that countries unable to exert much influence over each others affairs, loss of legitimacy by state.
Production and trade	Rapid technological change in industry and agriculture. Gradual shift of emphasis away from capital goods. Incentives for invention and innovation.	Import of Western technology and industrial models, large scale integration of enterprises. Continuing underutilisation of capital and labour. Conflicts over investment priorities between sectors.	Shift to profit rather than material output indicators of enterprise performance, shift to syndicalist enterprise management. Less stability of employment, possible increase in cyclical economic tendencies, increased difficulty in planning.	Attempt to combine humanisation of work with technological change, local self-management with nationally integrated framework. Some privatisation of agriculture possible.

Figure 4.3: *Continued*

	Scientific revolution	*Stagnation*	*Reform*	*Democratisation*
Distribution and consumption	Slow but smooth increase in consumption levels, fairly low inequality, considerable emphasis on collective services prices favour basic and collective goods.	Low consumption levels, resistance growing to poor quality products and increased prices of basic goods.	Rapid increase of material consumption, but involving occupational and regional inequalities. Possibly, extended welfare services and new forms of state provision.	Increased basic goods consumption, and reduction of inequalities in delivery of basic services. Improved mass communications and transport.
Political structures	Some liberalisation within framework of one-party state and its mass organisations and cultural institutions. Reduction of coercive activities of state, some increase in participation in enterprise management and political fora.	Growing political alienation and continued formation of small dissident groups. Consolidation of repressive apparatus, but no large scale return to terror, though advanced militarisation of society likely.	Growth of pluralism in institutions, more voluntary association, experimental pressure groups and lifestyle – but constrained by institutionalisation of management power, and use of coercion as means of economic control.	Use of existing institutions, but also reintroduction of pressure groups within parties, worker councils, and other means of achieving wider participation and popular scrutiny of state and management activity. Cultural revolution to overcome atomisation of society.

Figure 4.3: *Continued*

	Scientific revolution	*Stagnation*	*Reform*	*Democratisation*
Major dilemmas	Changes in working practices associated with increased productivity, and slow rate of expansion of consumption likely to create conflict between elites and mass. Strain likely in adding real content to hollow political framework, in overcoming engrained political alienation.	Political crisis has to be contained. Trade dependence on West liable to involve debt problems. Power of military may encourage adventurist policy, may provoke strains with other goup 3 countries.	Conflict between market-based economic nationalisation and desire for cultural openness, since the latter might permit expression of grievances about inequality, job loss, etc. Threatens powerful interests in state, e.g. military.	Different rate of change in different countries threatens stability of bloc, especially in face of real or imagined attempts from West to influence events. Demands for change likely to be unfulfilled in face of material problems.

(1) *Military–authoritarianism*: an extension of the neo-colonial political order with a military govenrment pursuing, in large part, the interests of transnational capital.
(2) *Developmentalism*: a development path pursued in support of national capital with considerable state intervention to this end.
(3) *Populism*: populist strategies of redistribution and import substitution pursued with much political mobilisation.
(4) *Socialism*: a strategy aimed at reducing external dependencies and mobilising workers to transform economic institutions.

These four scenarios, outlined in more detail in Figure 4.4, reflect the difficulties of Group 4 countries in the world economy. Each is rather precarious, as past experience has indicated: thus, populism has often been displaced by military–authoritarianism which in turn faces strains as it fails to satisfy the demands of many of its initial adherents. External support or intervention has played a critical role in the past, and there is little reason to expect its importance to cease. Thus, the global scenario will be extremely relevant to the long-term viability of these national scenarios.

Group 5 countries
It seems likely that oil prices will rise slowly, and stabilise at a high level, unless the world economy either declines or recovers dramatically. Thus Group 5 countries are likely to remain unusually prosperous, given that they are societies that are, in many respects, underdeveloped. We here focus on the most prosperous cases, those in the Near East.

Because of their entanglement in the region's problems and the common Arab culture that most of these countries share, the national scenarios necessarily contain a large regional element. One possible future involves much closer integration, although there are intense rivalries between the current dominant coalitions in the different countries. The scenarios we outline for this region are:

(1) *Western-oriented development*: this involves the pursuit of export-oriented industialisation, together with some westernisation of élite groups and cultures.
(2) *State-capitalism*: industrialisation strategy pursued under a high degree of state control towards heavy industries and infrastructure.
(3) *Neo-traditionalism*: radical nationalist movement, exercises power against foreign interests, with very interventionist and extensive state activity.

Figure 4.4: Scenarios for group 4 countries.

	Military-Authoritarian	Developmentalism	Populism	Socialism
Main Proponents	Transnational capital and local representatives (compradors), military and repressive state agencies, financial interests.	National industrial and agricultural capital, together with state agencies.	Large national capital, popular democratic movements, sections of workface.	Working class organisations, especially industrial workers, intellectuals.
Conditions of realisation of programme	Failure of populist projects, fragmentation of popular movements perhaps under influence of economic crisis and foreign pressure.	Discontent with inequalities, demand for state intervention, possibly results from failure of export-oriented policies during world recession.	Divergence of interest between local and transnational capital, deminance of large national capital over smaller firms, victory of populist leaders in elections.	Insurgent, mass membership, and/or charismatic mass movements intervene in national politics, linking industrial workers with urban unemployed and rural poor.
Production and trade	Lowering of wages, increased concentration of firms and reliance on foreign technology. Agribusiness developed. State provides infrastructure. Exports focussed on primary products, and some specialised luxury manufacturers. Some capital goods exports to poorer countries. High consumer goods imports.	Some nationalisation of foreign and some local assets, although they still operate with much autonomy according to market criteria and likely reinvolvement of foreign capital. Increase in extractive industries, partial planning of industrial development. Import substitution policies.	Nationalisation of transnational capital and some capital goods industries, agricultural land reform. Import substitution, less emphasis on exports, increased wages for industrial workers, less improvement for other groups. Public works maintain employment levels.	Effort to increase basic good production and reduce external dependency, and utilise labour and other local resources. Production more centrally planned, through markets remain important. Development of heavy industries and their internal linkages, search for long-term trade agreements benefitting backward regions.

Figure 4.4: *Continued*

	Military-Authoritarian	*Developmentalism*	*Populism*	*Socialism*
Distribution and consumption	Increased income inequality, stagnation of basic goods consumption while high-income groups consume more luxuries. Rapidly increasing urbanisation (perhaps being controlled by draconian policies of slum clearance). Growing regional inequality.	Initial effort to increase mass consumption. Taxation of luxury goods, but not to such an extent as to seriously depress consumption. No major shifts in wage differentials.	Attempt to shift effective demand to middle-income groups to restrict imported consumption habits. Public health and transport services increased. Expanded consumption of cheaper durables, services and basic goods.	Some steps to collective provision of basic goods, and growth of collective services with effort to reduce regional inequalities. Restriction of luxury consumption, some measure of income distribution.
Political structures	Military leadership make major state appointments. Repression of popular movements, attempts to use media and education to create national-authoritarian ideology, but possibly gradual moves to create managed parliamentary structures with limited party participation.	Essentially one-party state, with youth and labour organisations led by party members. Some repression of isolated attempts at contesting local officials or managers. Development of nationalist/religious/socialist rhetoric.	Likelihood of growth of technocratic groups in state (oriented to large capital); executive autonomy high perhaps with charismatic leaderships, but preservation of parliamentary forms. Nominally independent unions (whose leaders are likely to be drawn close to state).	Mobilisation of popular movements will determine degree to which democratic institutions remain active or become shells for power of party functionaries. Central issue will be participation of a broad range of social movements, together with decentralisation and internal democratisation of civil service.

Figure 4.4: *Continued*

	Military-Authoritarian	Developmentalism	Populism	Socialism
Major Dilemmas	Stagnation of major economic sectors, growth of large numbers of discontented marginal groups, rural hardship and food crisis.	Conflicts over pricing likely to become directly political. Risks of corruption and clientalism among leadership.	Dangers of currency crisis and inflation, of alienation of declining middle class and agricultural groups. Possibilities of corruption and restoration of privileges to state officials and other professionals.	Erosion of democracy into authoritarian forms, or of planning into state capitalism; foreign intervention: internal conflict among groups claiming to represent revolutionary ideals.

Figure 4.5: *Scenarios for group 5 countries*

	Western-Oriented	State Capitalism	Neo-Traditionalism	Endogenous Socialism
Main proponents or programme	National industrial and landowning classes, foreign capital.	National industrial and landowning classes and their representatives in the state.	Small national capital and petite bourgeois groups (e.g. merchants traditional leaders (e.g. of religious groups)	Urban and rural working classes, some intellectual groups.
Conditions of realisation	Western countries make concessions concerning regional issues (e.g. Israel), access to markets, and technology transfers. Defeat of radical movements by alliance of dominant groups.	Likely option for some countries in event of OPEC disunity, possibly involving political alignment with Second World. Weakening of landowner power, diversion of popular movements into national unity programme based on state power.	Polarisation between dominant groups (e.g. elite families) and wide range of popular movements, with increased ideological conflict and rejection of Westernisation. Mobilisation of ethnic or cultural groups to oust elements of dominant coalition.	Crisis of both existing and putative elites, weakening of military authority, emergence of strong leadership in unions and other workers' organisations. Facilitated by parallel developments in other countries of this and other groups.

Figure 4.5: *Continued*

	Western-Oriented	State Capitalism	Neo-Traditionalism	Endogenous Socialism
Production and trade	Move from import-substitution to export-oriented industrialisation, requiring considerable foreign support to overcome 'bottlenecks'. Perhaps some specialisation in particular industries (especially for small countries, and for petrochemical-based industries). Little agricultural development, large food imports.	Reform of agrarian structures, possible development of agricultural collectives and efforts to achieve greater food self-sufficiency (even exports). Aim at rapid industrial development with heavy industrialisation. Oil exports to Second World traded against technical assistance.	Combination of state capitalism and smaller-scale private enterprise. Expropriation of some transnational capital, at least in the first instance. Uneven attempts at reform and agricultural development, workplace decentralisation and recomposition of management structures. Some effort at self-reliance.	Land reform and efforts to provide suitable agricultural infrastructure (irrigation, machinery, and fertiliser production). Oil production limited to levels necessary to finance industrialisation for local needs. Emphasis on regional cooperation.

Figure 4.5: *Continued*

	Western-Oriented	State Capitalism	Neo-Traditionalism	Endogenous Socialism
Distribution and consumption	Regional transport systems developed to aid exports, using high technology. Relatively low living standards of masses, but these high relative to those of continuing pool of migrant labour. Imported Western consumer durables and luxuries: may lead to conflicts between rich and urban poor (who are likely to increase with migration from rural areas).	Continued reliance on migrant workers and foreign specialists, but rather less differentiation with living standards of masses. Some equalisation of consumption of collective goods. Mass education and literacy drives, with strong political overtones.	Criticism of imported consumption and welfare patterns, efforts to reintroduce lifestyles, housing, education, diet, etc. drawing on traditional forms. Some equalisation of consumption around relatively low levels. Employment maintained by public works, expulsion of migrant labour.	Emphasis on basic goods and collective services, but some reduction in living standards associated with reduction of consumer imports.

Figure 4.5: *Continued*

	Western-Oriented	State Capitalism	Neo-Traditionalism	Endogenous Socialism
Political structures	Military or traditional rulers supplement their coercive control with clientalism and establishment of official parties. Attempts to ward off conflict by large-scale public expenditure and an ideology of modernisation.	Possible tactical use of oil exports to West. Conflict between consumption aspirations of state officials and masses. Effort to integrate popular movements into state structures, or else to marginalise recalcitrant groups.	Conflict between different institutional power bases, probably being consolidated around a single, highly ideological leadership. State first promotes, then restrains numerous quasiautonomous cultural and political 'vigilante' groups. Dissidents and 'foreign agents' blamed for planning failures.	Revolutionary government installed, original aim to maintain high level of popular mobilisation. Evolution depends on internal developments (strength of labour movements, experience of dual power) and external factors (foreign intervention, trade prospects).
Major dilemmas	Uneven development and inequality, together with disturbance of traditional local power relations, promote cultural nationalism and/or high demands for increased consumption.	Elites continue to aspire to Western living patterns. Conflict with traditionalist groups (e.g. religions) and hierarchical workplace relations.	Likely to conflict with regional cultures and to provoke separatist currents. Interests of petite bourgeoisie and workers diverge and may lead to economic chaos. Fundamentalism may lead to high levels of internal repression and efforts to promote new model in other countries.	Possible drift of state toward bureaucratic one party repressive system or, faced with economic crisis, to state capitalism. External attempts to destabilise regime likely.

Figure 4.6: *Scenarios for group 6 countries*

	Neo-Colonialism	Liberal Developmentalism	State Capitalism	Bureaucratic Planning	Socialist Construction
Main proponents	Landowning oligarchy, comprador capitalists, foreign capital (e.g. plantation or mine owners/controllers).	Urban workers, some foreign capital.	Petite bourgeois nationalist groups urban workers.	Urban and rural workers and peasants. Technocratic state officials, some junior military.	Urban and rural workers and peasants, national liberation movements.
Conditions of realisation of programme	More difficult to assist if country very poor and/or dependent on a single export. Weakness and internal divisions of working classes; few international obstacles to Western domination.	Success of strategy limited without expansion of fairly restricted Northern markets. Industrial groups have to gain dominance over landowners.	Nationalist groups attempt to control more of a surplus through state intervention, legitimised by popular resistance to landowning oligarchies.	Displacement of capital from dominant coalitions, support of Second World countries, limited Western intervention. May involve radicalisation of military.	Leadership of new dominant coalitions needs to maintain close links – perhaps forged in liberation struggle – to workers and peasant movements.

	Neo-Colonialism	Liberal Developmentalism	State Capitalism	Bureaucratic Planning	Socialist Construction
Production and trade	Poor working conditions (including much exploitation of children). Limited industrial development, except for raw materials export. Enclaves of mining or other industry with relatively high consumption levels. Agricultural production becomes more market-oriented and capitalistic, with rural unemployment, but rural poverty limits productivity.	Some state intervention: provision of support for dynamic national sectors, regulation of transnational firms, land reform measures (beneficial to those richer peasants with some access to resources). Import-substitution policies, possibly quite inefficient and dependent on foreign technology and materials.	State control of profitable sectors (e.g. tourism, extractive industries, luxury foodstuffs). Continued dependence on technology imports, though attempts at import substitution and local production of capital goods. Land reform and perhaps use of cooperatives in agriculture though private large-scale agriculture may be retained	Effort to integrate agriculture and industry through production of capital goods for agriculture (may be undercut by imports from socialist bloc). Planning emphasises capital goods, and development of land and technical resources. High productivity demanded through moral and cash incentives, and public works used to limit unemployment. Collective agriculture, but private plots encouraged.	Effort to build on traditional technologies to maintain labour-intensive production and resist imported division of labour. Rapid development of technical education. Aim at decentralised capital goods production, related to regional needs, especially for agriculture and construction.

103

Figure 4.6: *Continued*

	Neo-Colonialism	Liberal Developmentalism	State Capitalism	Bureaucratic Planning	Socialist Construction
Distribution and consumption	Improverished peasantry, subject to famine in crop failures. Some urban workers and state functionaries well rewarded, as are traditional elites who help organise the system. Low wages and urban poverty, with few collective resources for other workers. Need for foreign aid.	High levels of regional and class inequality. Urban groups strive to follow imported consumption patterns, especially conspicuous consumption from middle classes. Some collective services for industrial workers. Low consumption in rural areas. High levels of rural and urban unemployment.	Some redistributive egalitarian measures, but differential rewards to skills mean rural-urban divergence maintained – especially if peasant incomes restricted to provide surplus. More consumption of basic and locally produced goods.	Attempt to promote regional and occupational equalisation, but these contradicted by need to maintain incentives for high production. Basic goods production increased, still largely distributed through market. Public works products, mass education and health campaigns, development of collective services.	Participatory planning and management encouraged. Effort to increase self-reliance in world economy. Emphasis on collective provision, satisfaction of basic needs, and limitation of inequalities. Local products consumed.

Figure 4.6: *Continued*

	Neo-Colonialism	*Liberal Developmentalism*	*State Capitalism*	*Bureaucratic Planning*	*Socialist Construction*
Political structures	Regime stability based on coercion and fragmentation of opposition, perhaps through exacerbating ethnic rivalries. Military rule frequent if not perpetual. Local landowners act as feudal lords in control of peasantry.	Strong concentration of political power, though political parties likely to exist, and some populist ideology likely to prevail if economic strategies successful. Depoliticisation of masses.	Single party regime likely, often seeking limited mobilisation through youth movements, militias, etc. State functionaries dominate, together with senior managers. Foreign relations present anti-imperialist rhetoric together with reliance on Western finance and aid.	Single party regime, reaching into local communities through cadre organisation of cultural education and village developed projects. Dangers of slippage from vigilance against foreign intervention to habitual intolerance of dissent.	Campaigns to improve economy and welfare used to maintain political mobilisation and education. Attempt to prevent bureaucratisation through widespread debate on current policies and leaderships.

Figure 4.6: *Continued*

	Neo-Colonialism	Liberal Developmentalism	State Capitalism	Bureaucratic Planning	Socialist Construction
Major dilemmas	Likelihood of growing social and ecological crisis, with endemic malnutrition and disease. Corruption of leaders liable to provoke outrage and continual challenge from lower strata of elites.	Conflict among elite groups, may reinforce regional or ethnic hostilities. Economic problems may require increased repression and austerity. Strong urban-rural divisions. May provide basis for neotraditionalist movements.	Planning goals imposed from to-down and liable to be rather arbitrary. Foreign capital liable to re-enter scene and change strategies.	Liable to be dissent from educated youth, workers facing production bottlenecks and slow increase in living standards. Efforts at destabilisation from outside interests.	Problems in shifting from current production patterns to new ones. Possibility of moving from 'self-reliance' to antagonism to all things foreign. Senior officials liable to develop own interests running counter to continued democratisation.

(4) *Endogenous socialism*: perhaps involving integration of several countries, reforming governments seek to bring about extensive redistribution of income and wealth.

These scenarios are outlined in Figure 4.5. Perhaps more so than for other regions, the course of events in one country is liable to influence that in others – especially for the microstates in the region. Group 5 countries in other regions may be viewed as resource-rich members of Groups 4 and 6.

Group 6 countries

We noted earlier that this group of countries is united more by poverty than by any positive feature, but that the range of national incomes contained in this group alone would be seen as vast, were there not far richer countries to compare them with. This means that countries pursuing similar development strategies may take very different times to achieve the logical outcome of these strategies (e.g. eradicating absolute poverty).

We shall treat Groups 6a and 6b together in the following exposition, as we do in the model. However, since the social structures of these subgroups have been diverging, their prospects for particular sorts of change are quite different.

Countries may move between the two subgroups as a result of internal developments or external pressure – and there is considerable likelihood of transition between scenarios grounded in market or planned economies. The first two of our scenarios are firmly within the sphere of Group 6a, the third scenario representing a possible outcome of either broad class of development strategy, and the last two correspond more to Group 6b:

(1) *Neo-colonialism*: continuing subordination of agricultural or mineral exporters, or geopolitically significant poor country, to economic or political interests of First World.
(2) *Liberal developmentalism*: growth of private national capital aiming at securing a hold in the world market; more generally feasible for well-endowed countries (e.g. minor oil producers).
(3) *State-capitalism*: national capital developed under state control of profitable sectors perhaps with populist ideology and social reform.
(4) *Bureaucratic planning*: attempt to follow Soviet model of central planning and heavy industrialisation.
(5) *Socialist construction*: attempt to originate culturally-specific mode of planned development aimed at satisfying basic needs and establishing new production relations.

These five scenarios are outlined in Figure 4.6.

CONCLUSION

We have surveyed what may seem a bewildering range of national strategies and scenarios, yet these are incredibly simplified compared to concrete circumstances and programmes of action in the contemporary world. By grounding our analysis in the interests of different social classes, and by drawing on the divergent experiences of countries where different dominant coalitions have formed, we have imposed some structure on this welter of contending forces.

The picture is, in part, very complicated because, as we have repeatedly stressed, national development policies do not exist in isolation. The viability of the various national scenarios depends very much on the global scenarios within which they are situated, as indicated in the descriptions above. In setting up our global economic model in Chapter 5 to 7, we shall take account of these interrelationships. Then, in Chapter 8 we shall build upon analyses to contruct a possible 'future history' for the world political economy up to the year 2000.

5 An Economic Model of World Income Distribution

THE REQUIREMENTS FOR THE MODEL

The previous chapters specified types of countries, social groups and relationships which need to be taken into account in a comparative evaluation of alternative development strategies. We may now begin to translate the findings of our scenario construction into operational terms, and describe a quantitative framework — a computer model of the world economy — with which to examine particular aspects of income distribution structures and policies.

The requirements for our model are twofold. First, it must be able to represent satisfactorily both the structure and the process whereby income distribution within and between different parts of the world are determined. Second, the variables in the model and the results which it generates must have a fair correspondence to those identified as crucial in the scenarios and worldviews we wish to explore. These two requirements are, however, incompatible, to some extent at least at the present state of the art in modelling. A thorough representation of income distribution requires a very detailed model. Models of income distribution for single countries (such as those of Taylor (1980) for Brazil, or Adelman and Robinson (1978) for South Korea), may describe ten or more income classes and sectors of production; while some global models include 'country models' for as many as fifty countries, but have only a rudimentary description of domestic distribution.

To build a model describing both international and intranational factors at this level of detail would be a Herculean task: to construct scenarios which assessed the likely behaviour of all the actors described by such a model would be incredibly demanding of resources. The only solution would be simply to assume that historical patterns of behaviour are just repeated in the future, or that whole blocks of actors behave together in a particular way. This

109

would defeat the purpose of the exercise. Consequently, our model should contain the minimum number of actors necessary to portray the phenomena we wish to explore.

The majority of global models have been directly concerned with the question of distribution between nations. With the exception of the Bariloche group's model, the economic descriptions in the earlier models largely ignore the question of redistribution *within* nations. The Bariloche group (Herrera *et al.*, 1976; Chichilnisky, 1977) demonstrated with their model that completely equal redistribution of *per capita* income within countries could have a large impact on the relief of poverty and malnutrition, and the satisfaction of basic needs for the entire world population. None of the global models addresses the vexed question of whether or not a drive for redistribution rather than growth provides the most effective way of achieving these ends, which is a major bone of contention between contrasting worldviews. Furthermore, several aspects of international distribution are disguised by the regional divisions used in the models. The problems of the poorest countries, or those with the most acute resource shortages, are hidden by regional averaging.

The results of any global model depend on the typology of countries used. and on assumptions about international trade and markets. technology. consumption patterns and population structures. Our model is no exception. We shall later provide a detailed description of our choices: here we summarise their most important characteristics. and compare them with the approach of other models.

First. let us consider the treatment of differences between economies. and the consequent typology of countries. Most global models group nations on a regional. geographic. basis (e.g. Latin America, South-East Asia). For our purposes it is more relevant to distinguish nations in terms of their level of industrialisation, economic and political structures. and potential for different kinds of development. We have already outlined an appropriate country classification in the scenario analysis. and this system is also employed in the modelling. Thus, at the very least, Third World countries should be divided into oil-exporting economies. the so-called 'newly-industrialised' economies. and the poorer economies. The last group, although numerous and varied. contains the lowest income nations, posing the more extreme challenges to present styles of global development.

Within the Northern industrialised economies we conventionally distinguish the centrally-planned economies from the market economies (i.e. the East and West). It is also useful to recognise that, among the market economies, some may be less ready to undertake innovations and complementary restructuring in the present crisis.

Indeed, some old industrial countries may conceivably be overtaken by the newly-industrialising economies and so be displaced from the dynamic centres of the world economy. For this reason we distinguish between two types of western economy.

A second issue we wish to represent in the model is the interrelation between these different economies and the way this affects income distribution *within* nations. Changes in technology or economic policy in one region may have substantial impacts — both negative and positive — on the rest of the world. New agricultural practices in wheat production in North America, for example, ultimately affect the livelihood of peasant farmers growing rice in southern India. To simulate such changes in our model, even in the most simple fashion, it must describe different kinds of technology and the chain of interaction between world and domestic markets.

It must highlight, too, certain relationships between the production structures and technology used in different countries, and the distribution of income and consumption within and between them. The use of a multi-economy, multi-actor model is very important here. The formalised theoretical results employed by proponents of the three worldviews (especially the conservative worldview), typically come from simple 'two-economy, two-agent' models. A growing body of literature exposes paradoxes in these models. With respect to transfers (e.g. aid) between countries, the simple models predict that the donor always loses by a transfer and that the recipient always gains (e.g. Samuelson, 1971). But the addition of a third agent (e.g. Gale, 1974; Chichilnisky, 1980) and the inclusion of substitution and production effects (Bhagwati, Brecher and Hatta, 1982) produces more ambiguous results — in some circumstances donors may gain and recipients lose.

Similarly, with respect to the sharing of benefits from international trade, some paradoxes arise. Some of these effects result from differences in consumption patterns, the imposition of tariffs, or substitution effects across producers (see e.g. Jones, 1982). Quite plausible differences in production techniques, or in the abundance of factors of production, can produce anomalous results even in the simple two-economy, two-country model (e.g. Chichilnisky and Cole, 1978a). Thus, since this model appears to be too simplified, the conventional wisdom that all parties gain from trade may itself be a paradox.

The third issue that the model should tackle, then, concerns production techniques and structures. Different production techniques have typically been treated in global models in terms of their characteristic demand for factors of production, for capital and labour. But different technologies require labour with different skills, so this description is insufficient. In assessing alternative development strategies, we need to explore this question. For example, we shall consider the implications for income distribution of adopting the appropriate techniques — so often advocated for the least industrialised economies — as well as the new labour-displacing technologies (e.g. microprocessor-related technologies), which may be the basis for a worldwide technological revolution; these technologies require extremely different skill mixes.

The fourth and final issue for the model is the description of households related to their income and patterns of consumption. Different types of household may be related to the different types of labour (more or less skilled) or financial resources which they may supply to the formal economy. What of different types of final consumption? According to Stewart and James (1982):

the choice of product is of major significance in its own right, independent of the consequences for factor proportions.... Product development in avanced countries designed to meet the needs of relatively rich consumers often produces 'inappropriate' products for poor countries.

Thus, as far as possible in a simple model, we want to explore the consequences of changing the composition of output as well as the techniques of production. To this end, we distinguish luxury consumption goods from basic consumption goods in the model: we can then represent the consumption of households with particular incomes in terms of mixes of these two types of goods.

One tool which enables us to incorporate the issue of distribution into our formal analysis is the Social Accounting Matrix (SAM). Such matrices are an extension of the more familiar input–output matrices used in economic accounting. They quantify the distribution of income, consumption and employment among different social groups — for example, rich and poor households, urban and rural workers and, in some cases, specific ethnic groups. The categorisation enabled by these SAMs provides an ideal structure for organising the data in our model.

We have outlined our approach to grouping national economies the division of factors of production (between labour and capital), and that of consumers (into high-income and low-income households). The rest of this chapter sets out our assumptions with respect to categories of production technology and human resources. Our discussion here will concern the *categories* of actors (nations, producers consumers), and Chapter 6 will describe the *behavioural* relationships between them. In Chapter 7, we shall use the model to explore the income distribution effects of the global strategies set out in Chapters 2–4.

THE SOCIAL ACCOUNTS FRAMEWORK

SAMs are a recent development in economic statistics, with which economists have sought to take account of the issues of income distribution at a national level. Until the 1960s, macroeconomic analysis formulated problems in highly aggregated terms, while microeconomic analysis concentrated on the behaviour of individual actors (e.g. firms). The focus of statistical agencies was on national economic accounts which gave information on key economic indicators, drawn up in aggregate, macroeconomic terms.

Over the last two decades, the need for more detailed data and forms of analysis has led to production data being disaggregated into many sectors, with financial flows separately treated, and with various classes of consumers identified so as to take advantage of the findings of more microeconomic studies. The proliferation of such statistics, and the fact that different categories are used to describe production, consumption and international trade, made the setting-up of a unifying system of national accounts imperative. In 1968, Stone devised a System of National Accounts (SNA) for the United Nations which enabled great variety and flexibility within a consistent, overall framework. (UNSO, 1968).

This new system of accounting forms the basis of the Social Accounting Matrix developed by Pyatt and Round (1979) in particular. Whereas the earlier input–output tables concentrated on the activities of production sectors, the SAM also showed the transactions of households and government. One advantage of this approach, is that the circular flow of income from factor demands to household income, to household demands for consumer goods, to production activities and back to factor demands, is explicit. The earlier input–output system only took interactions between production factors into account, so that such effects as the creation of indirect employment (i.e. new employment in one sector of the economy resulting from increased activity in another sector) could be badly misrepresented.

A second advantage for our purposes is that the SAM treats the distribution of income and consumption between households (or other institutions — e.g. government). Also, production activities within economic sectors can be subdivided between different kinds of production (e.g. urban versus rural manufacturing; modern versus traditional technology). Production activities can also be reclassified: in our model we have transformed categories of goods from the conventional definition in terms of production sectors (i.e. agriculture, industry, services) to a definition derived from the consumption patterns of households with different levels of income (i.e. basic, luxury and capital goods). Although the actual categories used in a given SAM may be deemed more or less important or appropriate by a particular worldview, these categories do not specify behavioural assumptions. For example, our distinction between different household types does not constitute a theory of the mechanisms of distribution of factor incomes. The accounting framework is thus useful for models grounded in rather different theoretical traditions.

Social Accounting Matrices have to date been prepared for about twenty countries. The layout of data in the different studies varies (see e.g. Round and Hayden, 1980). These accounts differ in terms of the number of production sectors identified, and the definition of household types. Some distinguish between urban-and rural-dwellers, for example — a distinction which is important for a developing economy with a large agricultural base, but of less consequence for an urbanised industrial economy. A number of problems become apparent when highly detailed SAMs are used for policy and distributional analysis (see e.g. Ward, 1981), one reason being that the conventional categories used to subdivide the matrix are not directly relevant to the theory of distribution or the policy being tested. A further problem (see Round, 1982; Hayden and Round, 1983) is that the elements of a highly detailed matrix, and consequently the multipliers derived from them, vary from year to year due to changes in the system of measurement or to fluctuations in behaviour; at higher levels of aggregation, these anomalies average out.

On balance, we choose to use a simple classification directly relevant to our theory. Even though some important distinctions are lost in a highly aggregated global model, they may sometimes be inferred from the classifications which are used. In some cases, the links are straightforward. (In particular, in the least developed countries, low-income households consume a high proportion of agricultural products. A large change in the level of basic goods production as defined in our model, therefore, would only imply a correspondingly large change in agriculture production, and thus in the level of rural employment.)

The layout of the data

The layout of data in the social accounts for our model, according to principal types of transaction between actors, is shown in Table 5.1. For each of the six regional economic groups we have located SAM data which can be organised in terms of this format.

Table 5.1: *Layout of Social Accounting Matrix*

	Producers	*Factors*	*Consumers*		*Totals*
Producers	Purchases of intermediate goods		Expenditure by rich and poor households Investment	Foreign trade	
Factors	Payments to skilled and unskilled labour and capital				Total Income
Consumers		Transfers to households from factors	Transfers between households	Transfers abroad	
	Total Expenditures				

Three major class of actors are identified: producers, factors of production and consumers. Each engages in a small number of transactions, indicated in the different segments of the matrix. Producers purchase intermediate goods from other producers (the top left-hand corner); they also make payments to workers (wages) and to capital (profits). No payments are made directly by producers to consumers as such, so there is no entry in the bottom left-hand corner. (In a more detailed matrix, this part of the table would show other costs to producers, such as indirect taxes on producers by government, and the depreciation of capital equipment.) In our model, only the overall impact of government is indicated, in terms of the transfer of income from high-to-low-income households through redistributive taxes and public expenditure, shown in the bottom right-hand corner of the matrix. The entry corresponding to depreciation of existing capital equipment is included in the top left-hand segment. The top right-hand segment of the matrix shows expenditures of domestic consumers on consumption and investment goods. Here we combine households, government and investors so that there is an implied association between these different categories, and rich and poor actors in the economy.

The six regional economic groups in the model trade and have other financial transactions (e.g. development aid) with each other. The net level of these transactions are shown on the right-hand side of the matrix. Just as we measure only the *net* transfer of income between high-and low-income households, we measure only the net level of exports (i.e. the differences between exports and imports for each good and the net income of households from abroad). If imports of a good are much higher than exports, then the economy has a high import dependence on those goods. Similarly, if total foreign income is much higher than transfers abroad then the economy has a balance of payments deficit.

The matrix is, therefore, divided in both directions into the three categories of actors: producers, factors of production, and consumers (including overseas consumers). The total expenditure of producers consists of their purchases from other producers, and the wages and profits they pay to factors. (In the SAM, the total expenditures are recorded in the bottom row of the table.) Similarly, the income of producers consist of sales to other producers, and purchasers by consumers at home and abroad. (The total income is shown in the right-hand column of the table). Total income and total expenditure are equal. The total income and expenditure of factors and consumers are recorded in the same way; and they too both balance their budgets.

For *producers* therefore we have the following accounts:

Total income   =   Sales of intermediate goods + sales to domestic consumers + net sales abroad

  =   Total expenditure

  =   Purchases of consumption and investment goods + net transfers to other consumers at home and overseas

For *Households* we have similarly:

Total income   =   Payments from factors of production + net transfers from other consumers at home and overseas

           =   Total expenditure

           =   Purchases of consumption and investment goods + net transfers to other consumers at home and overseas

For factors of production a similar budget equation holds, and, finally, there is a balanced budget for overseas transactions. If there is a net deficit in the balance of trade (i.e. imports are greater than exports so that net exports are negative), then there is an equal net financial transfer (which combines development aid, loans, interest repayments, repatriated profits). There is such a set of accounts for every group of economies: the total of net exports across the world for each commodity, and the total balance of payments, must be zero. Worldwide production and consumption are equal, and the world as a whole spends as much as it earns.

The SAM represents a set of transactions between activities. Although these activities are associated with actors, people may take on the roles of different actors, and engage in several or all of the activities of work, consumption and investment. A producer may also be a source of capital (an investor) and a member of a consuming household. Individuals are engaged in multiple roles, the relative strengths of which are measured by the elements of the matrix. The interrelationship between these multiple roles becomes more evident when the economic behaviour of actors is considered. For example, the trade-off made by households between their role as consumers and investors (i.e. capitalists) is indicated by their rate of investment.

In the model, each of the categories indicated in the table is further subdivided — production activities into basic, luxury and investment production; factors into skilled and unskilled labour and capital; and consumers into high-and low-income households.

We associate low-income households with basic consumption goods. For low-skilled workers, we trace the link from low skill to low wages, and hence, a consumption basket comprising mainly basic goods, and a relatively low level of saving and investment. This implies a low level of capital ownership (and thus little unearned income). Basic goods consumption requires basic goods production, which leads to a demand for unskilled labour. These linkages, shown in Figure 5.1, suggest a cyclical flow of income which under some circumstances may lead to a reinforcing, multiplicative impact on the income of poorer households – and in others to a vicious circle of poverty.

Figure 5.1: *The cyclical flows of income for low-income actors*

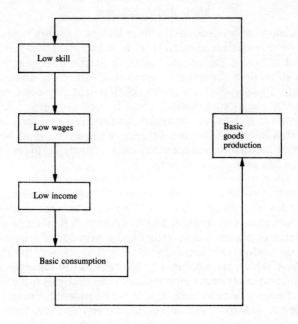

A corresponding and contrasting set of links for high-income groups is shown in Figure 5.2. Here, high skills are associated with high wages, and hence consumption of luxury and investment goods, both of which are more skill-intensive and capital-intensive in production. The figures imply two distinctive and independent self-reproducing systems; in reality, however, in most countries, both high-and low-income groups consume both luxury and basic goods, although in different proportions. (The precise proportions will depend upon the definition adopted for such categories as 'basic goods', 'skilled labour' and 'high-income households': we consider these later.)

Figure 5.2: *The cyclical flow of income for high-income actors*

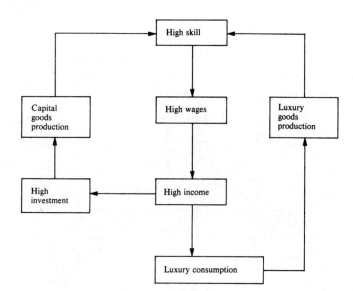

Both income groups also consume goods indirectly. For example, basic products are intermediate inputs in the production of luxury goods and capital goods, and capital goods are used in the production of all goods, incuding basic goods. These important interactions are given by the input–output matrix (the top left-hand elements in Table 5.1). Similarly, the production of all goods requires both high-and low-skilled labour, so low-income households gain income from the consumption of luxury goods by high-income households.

The picture is also complicated by foreign trade and financial transfers. For example, if most capital goods are purchased abroad, as is the case in the least industrialised countries, then the left-hand loop in Figure 5.1 will not function. These additional transactions are introduced in Figure 5.3, which shows the combined feedback between production activities, factors of production and consumers. The relative strength of the new transactions compared with those in Figures 5.1 and 5.2, has considerable impact on the pattern of economic growth. For example, an emphasis on the production of basic goods for domestic consumption would lead, over time, to a quite different pattern of income and wealth distribution between

Figure 5.3: *Interactions between actors in the domestic economies*

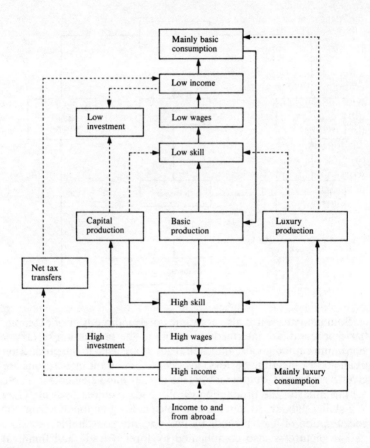

households than would one on the production of luxury goods for export.

There is, then, considerable room for debate among worldviews concerning the effects of development strategy on equality. A recent review of the empirical relationship between growth and distribution concludes that:

It is arguable, though far from proven, that a distribution-oriented development programme that integrates the poor into the mainstream of the economy may *cause* a higher rate of growth.... In the present state of our knowledge, we do not understand the economic, political or social dynamics of growth well enough to evaluate the merits of this argument. Research on the matter merits the highest priority among development economists and planners. (Fields, 1980, p. 241).

In other global economic models (see Cole, 1977) where the categories of household and workers are not differentiated, the idea is embodied that the income of all household groups changes in unison with national economic growth. A model which takes account of the differences in income-flows just described may contradict this assumption in many circumstances. In some countries, the conventional wisdom may apply: overall economic growth may be to the advantage of all economic actors. But in others it may be associated with worsening income distribution to the extent that some groups become considerably better-off, while others become relatively, and even absolutely, worse-off. The experiments with the model in Chapter 7 will illustrate this.

THE SUBDIVISION OF THE SOCIAL ACCOUNTS

The further division of the three categories of actor is best indicated by an example of the SAM used in the model. To illustrate, we take the accounts for the least industrialised Group 6 countries for the year 1975, where differences between high-and low-income households in consumption, and high and low skills in production are particularly evident. The accounts for the other economic regions will be considered later.

The social accounts for Group 6 economies are shown in Table 5.2. We shall not discuss the sources of the data at this point, but concentrate on how the social accounts operate. Each entry represents a transaction, with billions of US dollars (at their 1975 value) as the standard monetary measure. The total payments to skilled labour, unskilled labour and capital, shown on the right-hand side of the matrix, are $42 billion, $151 billion and $91 billion, respectively. The total wage income of skilled labour is, therefore, less than one-third of that of low-skilled workers. Since, as we shall see later, the low-skilled group make up 96 per cent of the working population, this means their average wage rates are *six times lower* than those of high-skilled workers. (All skilled wage income is assumed to go to high-income households, and all income from low-skilled labour to low-income households.) Payments from factors to households (bottom centre) show that the major part of profit (income to capital) goes to high-income households.

Table 5.2: *Example of a Social Accounting Matrix social accounts for least industrialised economies (US $bn 1975)*

	Production sectors			Factors of production			Final demand			Total
	Luxury	Basic	Invest.	Skilled	Unskill.	Capital	High-inc.	Low-inc.	Foreign	
Luxury	8	30	9	0	0	0	14	46	−11	96
Basic	39	161	36	0	0	0	30	160	6	431
Investment	11	35	12	0	0	0	41	18	−20	98
Skilled	6	17	20	0	0	0	0	0	0	42
Unskilled	17	127	7	0	0	0	0	0	0	151
Capital	14	62	14	0	0	0	0	0	0	91
High income	0	0	0	42	0	64	0	0	17	123
Low income	0	0	0	0	151	27	38	0	7	223
Total	96	431	98	42	151	91	123	223	0	1255

Note: Rows and columns may not add to precise total listed, in this and subsequent SAMs, due to rounding of amounts to nearest billion.

Production in these economies is largely that of *basic goods*. (This concept will be described in more detail later.) Basic goods production totals some $431 billion, compared with $98 billion for luxury goods and $70 billion for investment goods – as shown in the top right-hand column (total sales or output) and the bottom left-hand column (total of payments and purchase) of the matrix. The entry for payments to factors suggests that production of these goods uses quite different amounts of the two types of labour and capital. In particular, some 60 per cent of factor cost in the manufacture of basic goods $127 out of a total of nearly $206) is low-skilled labour. In the manufacture of luxury goods the corresponding figure is around 46 per cent, and in the manufacture of investment goods, only 17 per cent. These data suggest that on the production side for this group of economies, at least, the simple portrait of Figure 5.2, may have some merit.

The top left-hand entries in the table show transactions between producers. A notable feature is the relatively high input of basic goods into all kinds of production, including the production of the final consumption goods purchased by households. The indirect income flows of Figure 5.3 are not negligible; intermediate consumption accounts for a considerable part ($\simeq$ 57 per cent) of total output. A net transfer of some $38 billion from high- to low-income households is shown (bottom right-hand entries) together with a net transfer from abroad to both types of household. These latter transfers compensate for a net deficit on the balance of trade. The top right-hand entries show the economy to be a net exporter of basic goods ($6 billion) but a larger net importer of both luxury and investment goods. Indeed, imports of investment goods comprise about 61 per cent of the total investment goods purchases of all households, and 20 per cent of total consumption (including depreciation) by producers.

High-income households consume a significantly greater proportion luxury goods relative to basic goods than do low-income households ($\simeq$ 47 per cent of total compared with 28 per cent for low-income groups). And given the far greater number of low-income households, high-income households have a much higher *per capita* level of basic goods consumption (about 5:1). These differences reflect our definitions of basic and luxury goods, and the data presented in preceding paragraphs are highly influenced by these definitions. When we compare data across world regions, even greater income differentials are seen.

PRODUCTION AND DISTRIBUTION: THE DATA USED IN THE MODEL

The principal problem that arises when using Social Accounting Matrices is that the data required are not available for most countries. Macroeconomic data, such as the gross domestic product (GDP) and international trade statistics, are generally available. But the full set of data — including the distribution of sectoral earnings, levels of employment by different classes of workers, the rate of savings by households of different incomes, and the distribution of wealth between households — is only available for a small number of countries (and for different years).

Our model divides the countries of the world into six groups, as indicated earlier, in terms of their *per capita* income, resource base, level of industrialisation, technological potential and political orientation. In order to calculate the total income of each group, we add together that for each country separately. To calculate the regional *per capita* income, we divide total income by the total population of the region. Since relatively reliable data for both GDP and population are available for all countries (from United Nations and World Bank sources), this presents few difficulties. Data are also provided for all countries on the division of national value-added, and the labourforce between agriculture, industry and services. Data on international trade (using a slightly different sectoral classification) are also available, although some discrepancies are apparent when the imports and exports between countries or groups of countries are compared. And data on total investment and government expenditures are to be found in United Nations statistics.

The greatest difficulties come in organising data on income distribution and wealth in a form suitable for the model, and in estimating the inputs of high- and low-skilled labour to different sectors of production. Even if we have a clear understanding of what is meant by 'skilled' in a global context, the appropriate data do not exist and we must use some surrogate measure based on available data such as those for formal educational achievement. A third difficulty concerns the definition of sectors of production and consumption as basic and luxury goods, which does not correspond to the conventional aggregation of economic sectors. We shall adopt a definition in which the poorest income groups in the world consume mainly basic goods, while the highest income groups consume mainly luxury goods. The underlying assumption is that in developing countries, in particular, there is an association between the type of goods produced, the way these goods are produced, and the people who consume them. In the least developed Group 6 countries, our studies show that low-income households tend to provide the unskilled labour which produces the basic goods those households consume. An example of this would be a peasant farming family who grow and mainly consume their own produce, with their own livestock and land. In a more limited sense, this is also true for some aspects of production and consumption in the more industrialised countries of Groups 1–3. Many mass-production goods and public services are produced by lower-skilled workers and consumed in poorer households. What matters for our model is the extent to which such features are reproduced at the economy-wide level. Data on production in terms of a luxury and basic definition are not available for any country, so this must be obtained by transforming existing data on the composition of consumption of goods by households with different income levels.

In order to estimate the regional situation with respect to income and wealth distribution among households, the type of technology (its factor inputs of high- and low-skilled labour and capital) and the structure of production (basic, luxury and investment goods), data for specific countries are generalised to the group level. For each of the six economic groups identified in the model, a country for which a satisfactory data set is available, or can be compiled, has been selected. The data from these countries are then reconciled with macroeconomic data for the whole region for the base year (using a procedure described later).

With so many approximations in the preparation of data, with data often of dubious quality and based on an extremely restricted sample of countries, and with information compressed into highly aggregated categories, we can only achieve a crude representation of the world economy. The results of the modelling work, therefore, cannot be regarded as forming any kind of precise estimate of the circumstances of social groups or regions. The point of the analysis is to document broad tendencies and structural features rather than to detail specific circumstances. Provided the results of experiments with the model depend more on its structure than on precise details of the initial data used in the model, we may be more confident in our use of the limited data that are available. Studies of changes in the size of matrix elements in social accounts from year to year (see Round, 1982) suggest that the more income classes or production sectors that are represented, the more unstable and inexplicable the changes are likely to be. Thus, even if the accounts were prepared at a more detailed level, we should not necessarily expect to gain more reliable

Figure 5.4: *Preparation of data for the model*

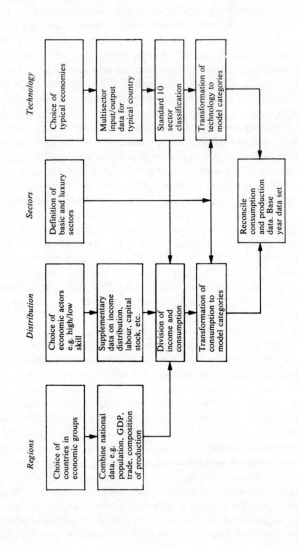

conclusions. There are definite limits to the degree of precision which can be reasonably expected from such models; ultimately, they can only be a guide to other micro-and macroeconomic analysis.

PREPARATION OF DATA FOR THE MODEL

Having described the general problems in the preparation of the social accounts, we now explain in more detail how the data were assembled and classified. In contrasting the magnitude of the entries in the table across the groups of economies, we shall see that they differ systematically between economies. In particular, the contrasts between the situation of high and low skill and income classes which we noted for the least industrialised countries, are less marked in other economies, especially the most industrialised — which, of course, is what the discussion in Chapter 2 would lead us to expect.

The preparation of the data involves four main tasks:

(i)   the allocation of countries into the six economic groups, and the compilation of appropriate data from statistical sources;
(ii)   the definition of economic actors (skilled and unskilled labour, capital, and high- and low-income households), and estimation of the allocation of income shares between them;
(iii)   the choice of countries to represent the technology involved in the production and consumption in each region, and the oganisation of data into a common format;
(iv)   the definition of basic and luxury goods and the transformation of the data assembled in (i), (ii) and (iii) into these categories.

The succession of tasks is shown in Figure 5.4.

GLOBAL CONTRASTS

The rationale for allocating countries into six groups on the basis of average income, present structure of production, technological potential and domestic and geopolitical orientation was described in Chapter 4. While this division is crude and leaves the situation of many countries ambiguous, there are very marked differences in the SAMs computed for the economic groups.

Table 5.3 shows how the average *per capita* income varies across the six groups. (These data are compiled directly from data in the World Bank Annual Report, 1980.) The richest group of economies has a

Table 5.3: *Population and average per capita income of economic groups*

Economy group	1	2	3	4	5	6
Domestic product (US$ bn 1975)	2277	1118	1087	542	244	284
Population (mn)	414	243	361	400	148	2468
Regional per capita income	5.5	4.6	3.0	1.36	1.64	0.11

combined income (i.e. gross regional product) of $2277 billions and a population of 414 million. The least industrialised group had a total income of $284 billion (about 5 per cent of total) and a total population of 2467 million (about 61 per cent of total). This represents a range of 48:1 in *per capita* incomes between average incomes of the richest and the poorest groups. (Even this figure does not take account of the wide variation in average *per capita* income of the economies in each group, discussed in Chapters 2–4, or the wide variation in income of different social groups within economies, which will be considered later).

The difference in average income between the least industrialised countries and other developing countries (the newly-industrialised countries and the oil-producing economies) is far greater than the differences between the latter and the industrialised economies. Even though the purchasing power of income varies across the world with a ratio of about 4:1 (according to studies by Kravis *et al.*, 1982), the data, nevertheless, confirm a very wide gap for the least industrialised countries.

These wide variations in the income and, hence, *per capita* consumption across economic groups, are matched by differences in the technology used in the different economies. One indicator of this is the average value of the capital goods used in production, *per capita*. Several estimates of capital stock for groups of countries in the world economy have been made (see Doblin, 1978), and from these we can calculate the total stock for each economic group, as shown in Table 5.4. More than 50 per cent of this capital, generating 41 per cent of the total world income, is located in the richest group of economies, who have only 10 per cent of the world population.

For the richest countries, capital stock *per capita* is $17,000, while in the poorest countries, the corresponding figure is only $318, giving a ratio of about 54:1. Capital stock *per capita* is one measure of the

Table 5.4: *Capital stock of economic groups*

Economy group	1	2	3	4	5	6
Total capital (US$ bn 1975)	7099	2875	1762	695	151	785
Capital/population	17.1	11.8	4.8	1.74	1.02	0.32
Capital/domestic product	3.12	2.57	1.62	1.28	0.62	2.76

global distribution of wealth, and appears to be slightly less equal than the distribution of income. We shall consider later what proportion of the population are actually employed in production. Taking these figures to indicate the capital versus labour intensity of production across the economic groups, suggests that the typical technology used varies across economic groups as much as levels of consumption. Capital intensity decreases from the richest to the poorest countries. In the Group 1 countries, $3 billion of capital are required for each $1 billion of income. The lagging industrial economies of Group 2 require about 80 per cent of this figure while in the newly-industrialised Group 5 countries, only about 40 per cent as much capital is used.

Given this trend for the richer economies, the figure for the least industrialised countries apears anomalous, although data given in Balassa (1982a, p. 1034) shows that the *incremental* capital output ratios for Singapore, South Korea and Taiwan (all newly-industrialised countries) are 1:8, 2:1, and 2:4, respectively, while for India the figure is 5:7. Part of the difference may be linked to the variations in purchasing power across countries, and to greater variations in the technology used within countries – or it may be that the level of capital stock is overestimated for these countries in Doblin's (1978) studies. Even so, the data suggest marked trends across economies, and confirm that, with respect to technology as well as *per capita* incomes, the very poorest group of countries differs distinctly from the other economic groups.

DISTRIBUTION OF INCOME AND WEALTH WITHIN ECONOMIES

The principal power relations in the world represented in the model are embodied in the distributions of income between labour and capital, of wage income between high- and low-skilled workers, and of wealth between high- and low-income households. Transfers between households, or between economies, also reflect power relations expressed in government policies.

Figure 5.5: *The allocation of income to major actors in the data preparation*

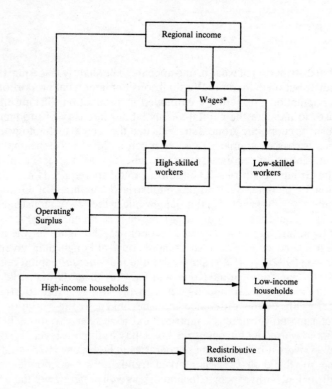

*Adjusted for indirect taxes

The distribution of national income among different types of household is estimated as shown in Figure 5.5. Total national income is first divided between profit and a wage income, making some adjustment for indirect taxes paid by production activities to government. Wage income is then divided between high- and low-skilled workers. Profit and wage incomes are then divided between high- and low-income households, with the redistributive effects of direct and indirect taxation accounted for. Although the procedure is straightforward, assembling the data is not. Only for very few industrial economies is a comprehensive and *consistent* set of data on both income and wealth distribution among households available in SAM form, and even here there are tremendous problems of definition and measurement. The data for income and wealth distribution have, therefore, been compiled from several sources.

The United Nations *Yearbook of National Accounts*, provides the data which enable a division of national income between wage income and operating surplus to be made for several economies in each group. These data are not presented for the Group 3, centrally-planned economies, and we derive these from a transformation of CMEA material accounts for Poland, by Kaminsky and Okolski (1981).

Table 5.5: *Operating surplus and rate of return for economy groups*

Economy group	1	2	3	4	5	6
Operating surplus % (before indirect taxes)	28	30	52	69	69	53
Operating surplus % (including indirect taxes)	24	26	52	33	33	32
Rate of return	7.8	10.0	31.8	26.0	53.8	11.6

The division of national income into operating surplus (i.e. profits) and wages before and after redistribution of indirect taxes, is given in Table 5.5. The data show that the share of profits in national income is typically twice as high in the developing economies as in the industrialised economies. Given figures for total capital stock by region, we can calculate the average return on capital (before capital gains taxes are deducted). As might be expected, by far the highest average rate of return is obtained in the oil-exporting, Group 5

World s Apart

economies — about twice that in the NICs and the centrally-planned economies, and four to six times that elsewhere.

The distribution of wage and profit income between skill and household types in each economy group is based on available information for economies in that group. For Group 2 countries, income and wealth distribution data for the United Kingdom are used exclusively. For other countries, a composite of data from a range of countries is employed. The first step in the preparation of these data is the construction of Lorenz distribution curves for before-and after-tax household distribution for each economy. (These curves show the proportion of each type of income which is received by a given proportion of the total population.) Thus, once we have specified the proportions of high- and low-income households in each economy, we may simply read their share of income directly from the Lorenz curves.

Figure 5.6: *Distribution of income and wealth for the least industrialised economies*

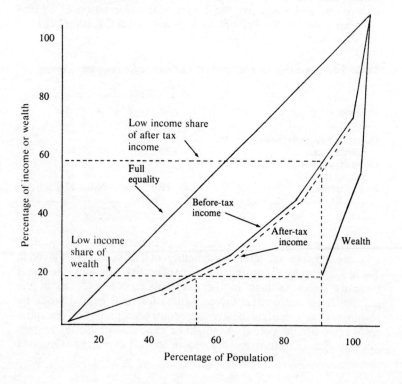

Figure 5.7: *Distribution of income and wealth for less dynamic industrial economies*

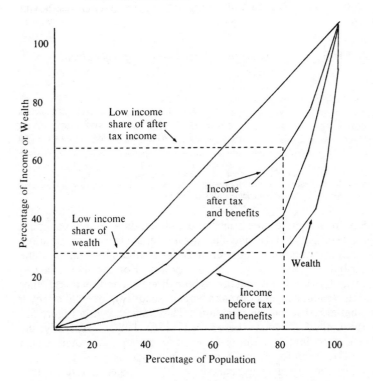

Figures 5.6 and 5.7 show the distribution of income and wealth taken for the less dynamic industrial economies (based on the UK) and the least industrialised countries (based on Egypt and India). Table 5.6 gives the share of total income received by the bottom 80 per cent of households in each economy group, and also for individual countries. As the discussion in Chapter 1 indicated, distribution of total income is significantly better for the most industrialised countries than for the developing countries, but that distribution is most equal in the centrally-planned industrial economies. Where data are available for several countries in each group, there is some variation, but the overall picture is maintained. For example, data for income distribution in Yugoslavia (World Bank, 1980) suggest a figure of 60 per cent of total income goes to the bottom 80 per cent of households, but according to data given in Kidron and Segal (1981) and Moroney

(1978), Yugoslavia, together with Poland and Hungary, has a worse income distribution than the Democratic Republic of Germany, Czechoslovakia, Romania and Bulgaria, but better than the Soviet Union.

Table 5.6: *Share of total income by bottom 80% of population*

Economy group	1	2	3	4	5	6
Average (1)	58	58	66	50	46	45
Typical (2)	57	61	66	46	42	50

(1) Shows the distribution of total income for the economic groups.
(2) Shows the distribution of income for typical economies; i.e. Federal Republic of Germany, United Kingdom, Poland, Philippines, Mexico and India.

Except for the more industrialised countries, data are not generally available on both before- and after-tax household income distribution; but such information is required in order to obtain estimates of the redistributive effects of government policy. For the industrialised countries the data show that taxes (both direct and indirect) and other benefits (such as pensions and social security), overall effect a substantial redistribution of income. These are shown in Figure 5.7 for the United Kingdom (based on HMSO, 1980). Generally, the level of redistributive taxation is greater for Group 2 economies (such as the UK) than for Group 1 economies (such as Japan).

Data for Hungary given in Szakolczai (1979) suggest that redistribution of income through the price system, together with the wider availability of services, is greater in the socialist economies than in the industrial market economies. However, extensive comparison is frustrated because the tax and transfer mechanisms differ between the two systems (e.g. Moroney, 1978). In general, the impact of fiscal and other policies on distribution is problematic, and very sensitive to assumptions concerning tax incidence. Because of this we take the redistributive element to be the balancing item in the budget equation of each household group in the model.

For the less industrialised countries, however, several studies report that the net effect on distribution of household taxes (both direct and indirect) is negligible. For example, Miller (1975) reports that taxation in Turkey has little impact on distribution. Foxley (1976) shows that in Chile the high progressiveness of direct taxation is almost completely compensated for by the regressiveness of indirect taxation; a similar conclusion is reached by Gupta (1975) for India. Thus, the shift in the Lorenz distribution for household income as a result of taking redistributive taxation into account, shown in Figure 5.6, is typically rather small in comparison with that for industrialised economies.

Figures 5.6 and 5.7 also show the Lorenz curves for the distribution of wealth. Data on wealth distribution are considerably weaker than for other sources of income. One estimation of wealth distribution for the less dynamic industrial countries is that of the Diamond Report (1977) for the United Kingdom (discussed in Hird and Irvine, 1979). This has formed the basis of estimates for the industrialised groups, the distribution of wealth being adjusted in line with the distribution of total income as suggested in Table 5.5. For the developing countries, the distribution of wealth has been based on Eckhaus et al's (1976) study of Egypt, and Pyatt and Round's (1977) of Iran. In making these estimates, the distribution of profit income is taken to be the same as the distribution of wealth. This implicitly assumes that the return on capital to all actors, and the return on capital across all production activities in each economy group, is the same. This is a questionable assumption, which we shall discuss further when considering the behavioural assumptions of the model.

The variety of sources from which our data have been drawn means that we must have reservations about the precise levels of distribution indicated, and the international comparisons implied. In the experiments we will be interested primarily in *changes* from these base year levels, and for this purpose the data probably form reasonable first approximations.

LUXURY AND BASIC CONSUMPTION

While our three economic sectors are not those commonly distinguished in international statistics, they do correspond to theoretical formulations ranging from the three departments of production of Marxian and neo-Ricardian economics to the sectors of neo-classical economic growth models. The description of consumption goods as basic or luxury is commonplace, and there have been many attempts to interpret the concept more formally. The idea of basic consumption can be used to describe the ideal of meeting all material, social, and even spiritual needs; or, more pragmatically, the bare minimum requirements of nutrition, shelter, clothing, and so on; or the trends of consumption that have been established as politically acceptable in particular contexts. Definitions of basic needs may be based on empirical as well as normative considerations. For example, the observation that poor people consume and spend a much higher proportion of their income on foodstuffs than do rich people is captured in the famous Engels' law.

An early formalisation of the idea of basic human needs by Maslow (1954) adopted a hierarchy of needs. The lowest were physiological needs for physical existence (e.g. nutrition) followed by physcial security needs (stability, freedom from fear), and so on. More elaborate descriptions extend such hierarchies and specify cultural and behavioural needs in more detail, in particular, ideas of pluralism, equity and social justice are included explicitly (see, for example, Lederer, 1980). On the other hand, empirical development research has tended to focus on very elementary lists of human requirements, such as those for nutrition and housing, where it is relatively easy to operationalise levels of need and progress towards satisfying them. Such research suggests a complex relationship between needs satisfaction and socioeconomic change: thus Hopkins and van der Hoeven (1983) could find no dominant variable – except possibly *per capita* GDP – to account for variations in countries' basic needs performance.

Whatever the philosophical starting-point, to incorporate the concept of basic consumption into a mathmatical model requires that it be dramatically simplified. In most models where the concept has been used, target levels for provision of material goods and social services are translated into a single *minimum income* target. Formally, it is implied, therefore, that recipients of this income will spend it in such a way as to satisfy their minimum needs, although the implementation of a basic needs policy inevitably requires social and other reforms. This was recognised, for example, in the Bariloche study (Herrera *et al.*, 1976); in the model, the raising of basic needs consumption was transformed into a maximisation of life-expectancy, but the authors also provided a programme of reforms needed to achieve the policies prescribed by their model (see Cole, 1977). Our use of the term 'human needs' in chapter 8 implies a similar assumption: that the strategy aimed at satisfying basic material needs requires social and political reforms as well as new economic policies.

In practice, most consumption goods have both a basic and a luxury component. How luxury or basic a given item is considered to be obviously depends on the economic and cultural character of a society. For example, expensive packaging, which makes up the major part of the final purchase price of a good, is not necessarily considered a luxury by affluent consumers. If it is not possible for consumers to purchase basic products without the luxury component (e.g. goods without expensive packaging) and, especially, if incomes are too low to enable the purchase of a sufficient amount of the necessary range of basic consumption items, then large proportions of all consumption will be a luxury. In this sense, too, goods with a high basic component will tend to be those consumed by low-income groups.

This idea that a particular good can be considered more or less basic derives from the concept of goods and services as consisting of bundles of characteristics. They are bought by consumers because of the desirability of these characteristics, and not as a result of some essence of the products themselves. If a characteristic helps to satisfy a basic need, it is classified as a basic characteristic; if it helps to satisfy some other need or desire, it is non-basic. Any product can, therefore, be divided into basic and non-basic components. A good is more basic than another if the ratio of its basic to non-basic component is higher. (Such a definition of basic to non-basic or luxury is closely related to that of Stewart (1978) and Stewart and James (1982), referred to in our introduction.)

Our model represents two key aspects of the basic needs concept. First, the consumption basket of low-income groups contains a higher proportion of basic goods than does that of high-income groups.

Figure 5.8: *The relationship between basic goods consumption and total per capita consumption*

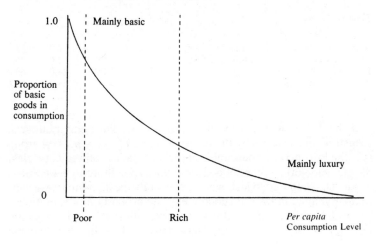

Second, it accepts that basic consumption is, in part, a relative affair, affected by historical choices. Poverty, welfare and deprivation are measured relative to other individuals and groups. The procedure with which the model captures these two facets is to assume that the richest group described in the model (i.e. the high-income households in the dynamic industrial economies) consumes a *low* proportion of basic goods and services, while the very poorest consume *only* basic items. Thus, the basic needs consumption of both income groups in all six economic groups is determined relative to the consumption of the other households. The exact relationship we employ is similar in form to that of Engels' law: the proportion of total consumption which is basic falls off exponentially as the *per capita* income of the household increases. (For example, we can assume that the proportion of basic goods in total consumption for a given class of household equals $\exp(-h/r)$, where h is the *per capita* expenditure on all consumption goods by those households, and r is the corresponding figure for the richest households identified in the model. This is shown schematically in Figure 5.8.)

Since households with different incomes consume different amounts of agricultural, manufactured goods and services, the concepts of basic and luxury goods may be applied usefully to each of these items separately.

Table 5.7: *Consumption of agricultural goods and services by households*

Economy Group	1	2	3	4	5	6
Rich Agriculture	3.5	3.6	9.5	5.0	9.0	35.0
Services	58.0	56.0	52.0	54.0	52.0	45.0
Poor Agriculture	5.0	5.4	11.0	24.0	23.0	60.0
Services	52.0	51.0	50.0	45.0	48.0	22.0

Table 5.7 shows the proportion of total expenditure on agricultural, industrial and services by the twelve groups. The very poorest group (i.e. the low-income households in the least industrialised countries) devote some 60 per cent of consumption expenditure to agricultural products, compared to just over 3 per cent by the high-income groups in the dynamic industrial economies. The proportion of agricultural goods diminishes very rapidly in favour of industrial goods consumption for incomes above the very lowest. For the high-income

households in the least industrial countries, expenditure on agricultural products is only 35 per cent, considerably less than for the lowest income households. Taking account of their very different total disposable income, actual *per capita* expenditure on agricultural products for the two groups, is $1987 per person annually for the richest group, and $460 for the poorest. The formula above suggests, therefore, that about 78 per cent of agricultural consumption by this lowest income group should be described as basic. Applying this rule to each commodity in turn for all households gives the aggregate proportion of expenditure on basic goods by household. This is shown in Table 5.8.

Table 5.8: *Basic consumption by households (%)*

Economy Group	1	2	3	4	5	6
Rich	13.5	14.1	28.8	33.7	49.3	68.0
Poor	33.7	40.5	54.9	69.0	61.4	77.7

LABOUR SKILLS AND INCOME DISTRIBUTION

High- and low-income households are directly identified in the model by whether they recieve wage income from skilled and unskilled labour. The technology used in the production of goods is described in terms of the use of capital and the two kinds of labour. Thus, the precise definition of skill we use will be central to the setting-up of the social accounts: it determines the distribution of income between labour and households which are to be displayed in the social accounts.

The definition of skill poses many questions. Although we simply divide production labour into the categories skilled and unskilled, a further division is essential for most analysis. But in the context of our present wider discussion this division is not unreasonable. Of the two main strands of research here, the occupational psychology approach (e.g. Crossman 1966; Hazlehurst *et al.*, 1969; US Department of Labour. 1970) and the occupation activity approach (e.g. Keesing, 1966: Hufbauer, 1970: Hallak, 1978), the latter suggests a fair correlation between economic performance and skill intensity of production at the sectoral level. This literature concentrates on the differentiation of production skills; certain occupational categories, such as scientists and engineers, and conceptual and discretionary competences may be more relevant to the idea of creative skills and hence to technological *change*.

In assembling data we have, as far as possible, defined skilled workers as people with higher education, together with people who have received equivalent vocational training: unskilled workers are all other people of working age. Relevant statistics for each region is calculated from data for individual nations given in the ILO *Yearbook*. Table 5.9 shows that the skill intensity of production declines as we move across the six groups of the economy described in the model. In the most industrialised economies, about 20 per cent of employees are skilled, falling to only 4 per cent in the least developed economies.

Table 5.9: *Proportion of skilled workers and share of total income*

Economy group	1	2	3	4	5	6
Skilled workers (%)	20	18	15	9	6	4
Relative wage rates of high-and low-skilled workers	2.1	3.2	2.0	3.2	9.2	6.8
Share of total income to high-skilled workers (%)	34	41	26	24	37	22

Based on *ILO Yearbook* (1975) and other sources (see text).

Given our restricted definition, a more skilled worker is likely to be more productive than a low-skilled one, and is paid a higher wage. Many studies justify this assertion. For example, Layard and Saigal (1966) found an association between occupational categories and labour productivity. They found also that an increase in the proportion of professional and technical workers in manufacturing industry was linked to an increase in productivity. Horowitz *et al.* (1966), and later studies by the OECD (1970), have suggested a similar relationship. The association between wage levels and educational attainment has also been demonstrated. Hinchcliffe (1975) shows that lifetime earnings of workers with higher education in the United States are about 2.5 times those of primary educated workers. In India the ratio is about 5.5:1, and in Nigeria, about 10:1. In general, the gap between different groups of workers seems to be higher in Third World countries (see data in ILO, 1978). Combining data on wage rates with the data on the composition of the workforce provides an estimate of the share of total wage income in each economy, which is earned high- and low-skilled workers, respectively. These estimates are shown also in Table 5.9.

In the most industrialised countries, about 34 per cent of wage income is received by skilled labour, which makes up about 20 per cent of the workforce. In the least industrialised economies the corresponding share of income is 22 per cent. This is the allocation used to divide wage income between high- and low-skilled workers in the least industrialised economies (i.e., in the appropriate SAM, Table 5.2, $42.5 and $150.7, respectively). These figures are problematic, not least because we are applying the same definition of skill across all countries. At best, therefore, we can only expect these data to be broadly indicative of cross-national trends.

Our definition of skill is limited in that it is concerned mainly with production skills. But even here there are several considerations necessary, especially with respect to developing economies. First, the supply of skills from the education sector is measured in terms of the number of years spent at school, formal qualifications, and so on. Since both schooling and qualifications in the post-colonial era have been based on Western standards, an assessment of *available* skills is likely to miss many indigenous skills, especially those communicated through informal education in the home and outside. The extent of the unaccounted skills is difficult to evaluate, although some effort has been made in this direction (e.g. Acero, 1981). Similarly, on the production side, demand for skills is related to the technology employed in the modern sector. Since the technology there is often derived from Western sources, skill requirements are measured in *similar terms*.

Second, production skills are usually thought of as those relevant to employment in the formal sector. Much less effort has been put into questions of employment or skills in the informal sector. In developing countries, the informal sector is closely related to the traditional sector, and, in industrial economies, to the so-called 'underground economy'. Estimates vary considerably, and also suggest that different countries have very different situations in this respect. For example, some estimates for the UK suggest that the informal economy accounts for 7.5 per cent (see Gershuny, 1979) of domestic product, and in the United States for 10 per cent (Guttman, 1977; Rutherford, 1978). In other industrial economies, (e.g. Italy and Poland) it may account for 20–5 per cent of all production, and even more in some developing economies.

HOUSEHOLD INCOME

The relationship between skill and the distribution of wage income
between households in the model is as follows: we assume that all
wage income from high-skilled labour goes to high-income households,
and all wage income from low-skilled workers goes to low-income
households. Thus, households are defined primarily in terms of the
labour skills they provide to the formal sector and can be thought of
families dependent upon high- and low-skilled labour.

The proportion of high- and low-income households in each
economy group is taken to be roughly the same as the proportion of
high- and low-skilled workers (Table 5.9). With these simplifying
assumptions we may calculate directly the share of household, wage
and profit income from the Lorenz curves and data on wage
differentials described earlier. These figures are given in Table 5.10,
which shows that the share of all income from all sources, and the total
after tax to high-income households, is lower in the developing
countries than in the industrialised countries. This is mainly due to the
fact that, with the definition we have chosen, there are relatively fewer
high-income households in the less industrialised countries.

Table 5.10: *Income shares to low-income households*

Economy group	1	2	3	4	5	6
Low-income households (%)	80	82	85	91	94	96
Share of wage income	66	59	74	76	63	78
Share of income from profits	22	25	22	35	35	30
Share of income after tax and benefits	57	58	58	64	73	76

Table 5.11: *Per capita consumption of households by income (US$ 000s 1975)*

Economy group	1	2	3	4	5	6
High	6.6	6.4	4.0	3.6	2.2	0.44
Low	3.5	2.9	1.8	1.5	1.1	0.087

Differences in *per capita* consumption follow those in income. Table 5.11 gives the *per capita* consumption for the twelve household groups included in the model. It shows that the *per capita* consumption of the richest group are some 72 times those for the poorest group.

PRODUCTION TECHNOLOGY

The estimation of production technology is based on data from typical countries for each economic group. Their factor inputs are used as a guide for the region as a whole. For example, if the chosen country's production of a given commodity is particularly intensive in the use of unskilled labour, this must be reflected in the final data set for the group. If the technologies used in the production of different goods between typical countries are very different, then this too must be reflected in the final estimation.

The procedure whereby this transformation of production and consumption categories is made is lengthy, so we do not elaborate fully here. The complications arise mainly because the many important differences between consumption of high- and low-income households are masked when sectors are highly aggregated. To take an example: the *type* of foodstuffs purchased by the various household groups may differ at least as much as the *amount*. Within a particular category of food (e.g. fresh meat) high-income households may consume grain-fed beef, while low-income, subsistence households consume goat meat fed on scraps and forage. Thus, production technologies for this good as consumed by the two households are very different. This phenomenon, too, is masked when working at a high level of aggregation in which many kinds of foodstuff are combined. As far as possible, with available data, we have tried to reduce this effect using the procedure indicated next. This was carried out for each region at the ten sector level, but, even with this detail, the differences we want to highlight were substantially masked.

The main steps in the calculation can be understood by comparing Table 5.12 with the SAM for the least industrialised Group 6 in Table 5.2. Table 5.12 shows the SAM for the least industrialised economies, when production and consumption are sub-divided into conventional agriculture, industry and services categories, and with government expenditures shown explicitly. This table is based on data for India, one of our typical economies (see Cole and Meagher, 1981). It was obtained by scaling the composition of value added by production sector and the composition of value added by factors of production, so

Table 5.12: *Social accounts for least industrialised developing economies (US $ bn 1975)*

	Production sectors			Factors of production			Final demand					Total
	Agric.	Indust.	Service	Skilled	Unskill.	Capital	High-inc.	Low-inc.	Govt.	Invest.	Foreign	
Agriculture	13	29	2	0	0	0	18	74	1	1	6	144
Industry	43	108	22	0	0	0	15	37	8	57	-28	262
Services	3	49	17	0	0	0	18	43	34	1	-3	163
Skilled	1	7	20	0	0	0	0	0	0	0	0	27
Unskilled	76	30	28	0	0	0	0	0	0	0	0	133
Capital	8	30	67	0	0	0	0	0	0	0	0	105
High-income	0	0	0	27	0	42	0	0	0	0	13	83
Low-income	0	0	0	0	133	28	0	0	0	0	6	167
Government	2	11	6	0	0	35	0	-1	19	0	6	77
Investment	0	0	0	0	0	0	32	14	14	0	0	59
Total	144	262	163	27	133	105	83	167	77	59	0	1220

Notes: as Table 5.2

that the Indian data correspond to the value added for the economic group as a whole.

Each matrix element is adjusted by the minimum amount necessary to ensure that the total income to each factor from all three sectors equals their share in total regional income. (This procedure is based on the technique developed by Stone for the SNA system mentioned earlier, and is described in detail in Meagher, 1980.) A similar procedure is used to scale the intermediate inputs to production. In this case, the rule is applied that the ratio of total intermediate consumption to value added by sector should be the same for the regional economy as it is for the typical country. These data are then transformed together into the luxury, basic and investment goods categories using the rules presented by Figure 5.8.

Other data, such as the distribution of profit income between households and the rates of taxation, are also adjusted to match the data presented in the previous section. Similarly, levels of trade are set to match data compiled from international sources. (Since this procedure is straightforward, we do not discuss it further here.) The overall adequacy of this calculation rests on how representative the 'typical' country is of the economic group as a whole. Given the very great diversity of countries in each group, and the limited amount of data from which to choose, a good deal of approximation is involved. Nevertheless, the variation in technologies between the economic groups is sufficiently wide that, the procedure provides useful profile of each economy.

Table 5.13: *Factor inputs for agricultural production*

Economy group	1	2	3	4	5	6
Labour/capital*	4	11	14	568	976	1563
Skilled/unskilled	0.059	0.053	0.017	0.003	0.011	0.002

* Workers per US$ mn 1975

The variation in technology across the various economy types is illustrated in Table 5.13, for all agricultural sectors combined. This shows that the ratio of capital to labour inputs varies by a factor of 390 between the least industrialised in Group 6 and the most industrial countries in Group 1. The variations are less marked for industrial goods and services, but the predominance of agricultural consumption for the lowest income groups makes this constrast in agricultural technologies especially significant for comparing basic goods production technology across economies. Within economic groups there are also very marked differences, especially in the least industrialised economies, where the ratio of capital to labour inputs differs between agriculture and services by a factor of 4.5. These data are in line with the findings of the OECD interfutures project (1979) and UNIDO (1978) study cited earlier.

Table 5.14: *Labour/capital ratio by sector (workers per US$ mn 1975)*

Economy group	1	2	3	4	5	6
Luxury	28	40	110	290	630	1130
Basic	28	41	90	310	540	1870
Investment	36	54	100	240	260	620

The resulting estimates of production technology for basic, luxury and investment goods are given in Tables 5.14 and 5.15. These show, respectively, the ratio of total labour (skilled plus unskilled) to capital (worker-years per $1000 of capital) by sector, and the ratio of skilled to unskilled workers employed by sector. The most notable features of these data are (i) that the production technologies for the different goods diverge considerably more in the developing countries than in the industrialised countries; and (ii) the proportions of skilled labour and capital used are less in all sectors in developing countries than the industrialised countries. The similarity of the production technology for basic and luxury goods in industrialised countries follows principally from the similar consumption patterns of high- and low-income households; the diversity in the least developed countries follows from the large differences in consumption patterns and the much greater

Table 5.15: *Skilled/unskilled ratios by sector*

Economy group	1	2	3	4	5	6
Luxury	0.26	0.24	0.15	0.12	0.057	0.053
Basic	0.23	0.19	0.17	0.08	0.056	0.020
Investment	0.25	0.21	0.25	0.11	0.112	0.450

heterogeneity of production technology across agriculture, industry and service sectors. Together, these features help to determine the strengths of the high and low cyclical income flows shown in Figures 5.1 and 5.2.

CONTRASTING PRODUCTION STRUCTURES, TRADE AND LEVELS OF DEPENDENCE

The previous sections have demonstrated the great heterogeneity of patterns of income distribution, consumption and production techniques across the economic groups. These data are summarised for all economic groups in Tables 5.16 and 5.17, which show the row totals of the Social Accounting Matrix (i.e. the total income of each category of production, factor and consumer) and levels of international trade. Only one adjustment has been made: the transfer from high- to low-income household has been deducted from the entry for high-income households. The Table 5.16 shows the value of items in US $ 1975. The Table 5.17 shows the percentage share of each commodity in total output, the share of income going to each factor or household, and the size of net exports relative to total production in each economy.

Table 5.16: *Summary social accounts for all economies: base year 1975 (US$ bn)*

Item	1	2	3	4	5	6	World
Output:							
Luxury	2756	1088	937	277	124	96	5278
Basic	1049	509	783	422	172	431	3367
Investment	919	431	424	170	84	98	2125
Factors:							
Skilled	586	339	137	87	60	42	1252
Unskilled	1138	488	390	275	103	151	2544
Capital	552	291	560	180	81	91	1755
Households:							
High-income	952	563	587	225	114	123	2565
Low-income	1280	650	634	358	178	223	3324
Net trade:							
Luxury	27	−12	−15	−16	27	−11	0
Basic	−25	−9	32	8	−13	6	0
Investment	81	13	−34	−25	−15	−20	0

Table 5.17: *Summary social accounts for all economies: Base year 1975*
(percentage of income in each category) (1)

Item	1	2	3	4	5	6	World
Output:							
Luxury	58	54	44	32	33	15	49
Basic	22	25	37	49	45	69	31
Invest.	19	21	20	20	22	16	20
Factors:							
Skilled	26	30	13	16	25	15	23
Unskilled	50	44	36	51	42	53	46
Capital	24	26	52	33	33	32	32
Households:							
High-inc.	43	46	48	39	39	36	44
Low-inc.	57	54	52	61	61	64	56
Net trade:							
Luxury	1	−1	−2	−6	22	−11	0
Basic	−2	−2	4	2	−8	1	0
Invest.	9	3	−8	−15	−18	−20	0

Data are percentages
(1) Net exports are percentage of output by sector

The world total for each entry is also shown. We should caution that while the description of goods and factors of production across regions are roughly equivalent across economies (at least at this high level of aggregation), this is hardly the case for household categories. In particular, although households are defined in terms of their ownership of skills, the fact that wage rates and production technologies vary so widely across economies means that the consumption patterns of households supplying similar skills are quite different across the world.

The first entries in the tables give the total output (i.e. final domestic consumption, intermediate consumption, net exports). Thus, basic goods production makes up 69 per cent of the total value of ouput in the least industrialised countries, but only 22 per cent in the richest economies. For the world as a whole, the figure is 31 per cent. Our data show that about 83 per cent of all investment goods production is in the industrialised world (Groups 1 – 3), and about half of developing country production of investment goods is in the newly-industrialised Group 4 economies. Studies by UNIDO (1978) and the OECD Interfutures Project (1979) show that over 98 per cent of capital goods production is in the industrialised countries, while nearly 80 per cent of production of the remainder is in only eight countries. Our data show a higher proportion of local production in the developing world, because we take account of all investment including infrastructure expenditure, as well as the purchases of machinery and transport equipment recorded in the UNIDO and OECD studies. Further, India and China are grouped with our least industrialised countries because of their low average *per capita* income and overall economic structure.

The contrasts in the composition of consumption across economies reflects both the composition of demand by domestic consumers and production sectors, and the relative factor endowments of the economies. The composition of demand for basic, luxury and investment goods is defined largely by the levels and distribution of income shown in the tables, and also by the availability of different goods. The shares of income accruing to skilled and unskilled labour and capital too are, in part, a consequence of the availability of these factors and access to technology. (Some of the factors which determine these variables, according to the different worldviews, have been mentioned in the discussions of Chapters 3 and 4.)

The bottom entries of Table 5.16 show the levels of net exports for goods as they appear in the SAM. The trade data for the model have been compiled from a number of sources. A trade matix for the economy groups was first prepared from data supplied by the United Nations (Kittler, 1979), and then modified to agree with the more explicit data on CMEA Group 3 economies reported elsewhere (GATT, 1978). All these data are reported in terms of the conventional trade categories (SITC), so, again, a transformation to the new categories of basic, luxury and capital goods used in our model is required (described in Evans and Cole, 1979; and Meagher, 1980).

The picture of dependence between the economic groups which these data provide is summarised by the net levels of exports for each commodity and type of economy. A large net import of a good, relative to the level of domestic consumption, suggests that an economy is especially dependent on that good. The data on investment goods trade show the dynamic industrialised economies of Group 1 to be the dominant net exporter, with Group 2 economies contributing a smaller share. All developing country groups are net importers of investment goods. For the least industrialised countries, the dependency ratio of imports to consumption is highest at about 20 per cent, with 18 per cent for the oil-exporting economies, and 15 per cent for the newly-industrialised economies. These ratios are shown in Table 5.17.

The data on capital goods trade reflect the relative weakness of the developing economies in their production of capital goods, given the relative skill-intensity of capital goods production in these countries (see Table 5.16), where skilled labour is relatively scarce (Table 5.9). By comparison with the developing economies, all the industrial groups (including the centrally- planned group) show a rough balance in their consumption and production of investment goods. Although these economies may be vitally dependent on *particular* equipment or commodities, their overall levels of trade for each group of commodities is small relative to their levels of trade for each group of commodities is small relative to their levels of output, and in this sense, they are less economically dependent.

The situation for international trade in luxury goods is similar to that for investment goods and for similar reasons. Again, the most industrialised economies dominate this aspect of world trade. The principal exceptions here are the net import of luxury goods by the less dynamic industrial economies and the net exports from the oil-exporting economies. (This results largely because oil energy is a luxury good according to the definition we have adopted. By contrast, firewood, as a staple but untraded source of domestic energy in the least industrialised countries, would be counted as a basic good.) All other countries are net importers of luxury goods.

By contrast, the most industrialised economies are net importers of basic goods despite the fact that the USA, the largest member of Group 1, is a major food exporter. The least industrialised Group 6 countries are net exporters of basic goods, despite their relative poverty; although basic goods are a large proportion of total consumption, absolute levels of consumption of even basic goods are very much lower than in other economies.

For all regions, there is a net imbalance on international trade, and all, except the most industrialised countries, show a deficit for the base year (1975). This applies also to the lagging industrial countries of Group 2, whose net imports of both luxury and basic consumption goods are not offset by their net exports of capital goods. (We shall say more about the balance of payments in Chapter 6, when we discuss the mechanisms of distribution in the domestic and international economy.) The requirement that each actor observes a balanced budget obviously means that any imbalance of trade must be offset by a corresponding transfer of aid, foreign investment and other payments.

This completes our discussion of the structure of the six economic groups represented in the model. We have emphasised the diversity in production technology, consumption and income distribution within and between economies and international trade between them. This diversity will be a major determinant of the results of the experiments with the model to be described in Chapter 7.

6 Mechanisms of Distribution and Change

In Chapter 5, we established a set of accounts for major groups of countries in the world, and commented on some significant aspects of their economic structures. But the discussion said rather little about underlying mechanisms which led to the present pattern of distribution, or which might bring about change. By contrast, the chapters on scenario building concentrated largely on explanations of such mechanisms and alternative strategies for global development. The statistical picture of distribution and consumption, so far presented, might be broadly acceptable to proponents of all worldviews, despite its limitations; however, there is much less agreement on the mechanisms of change. Theoretical controversies here include, in particular, the determinants of consumption and saving by different income groups, the sharing of surplus between different factors of production, the mechanisms for technical change, and the determinants of international trade and financial flows.

There is no way to find a lowest common denominator of these theories. But we can select a particular set of behavioural relationships which we consider to be key to the development strategies outlined in Chapters 3 and 4 and use these as a starting-point for further discussion through the experiments with the model described in Chapter 7.

WORLD MARKETS AND INTERNATIONAL FINANCE

The interaction between domestic and international markets is clearly a vital concern for each worldview. The conservative worldview emphasises the beneficial effects of free trade, while the reformist and radical views (especially the dependency theorists) point to the detrimental consequences of the operation of world markets in practice. International markets have become an increasing force in the world economy, and the volume of world trade expanded much faster than domestic production over the post-war economic boom. One important issue to explore, therefore, is how the operation of world markets affects the distribution of production and consumption worldwide. We shall do this in a relatively simple fashion, and assume that all goods produced in the world are tradeable, and that as a result there is a 'world price' for each type of commodity. Thus, when we carry out an experiment with the model to determine, for example, the results of an increase in the level of development aid, we allow the world economy to adjust until a new set of world prices is achieved.

In effect, we are implying that the world economy moves from one equilibrium to another, and that there is a worldwide system of production and consumption which eventually settles down after any new policy or change in behaviour. This is obviously an exaggeration since, in reality, the categories of goods we are dealing with are very heterogenous. Some cannot be traded internationally, while for others there are distinct markets with different prices for very similar commodities. Much trade, for example, is carried out between subsidiaries of international firms, who may adjust their prices in order to redirect profits into countries with favourable tax arrangements. Data assembled by Nayyar (1978a), Helleiner (1980) and Lall (1980) suggest that typically between 10 and 50 per cent of international trade by newly-industrialising countries is via international firms. In some cases, embargoes are placed on stratiegic materials, restricting trade absolutely, while most other trade is affected by quota or tariff systems. Consequently, the post-war reforms signalled by the Bretton Woods and GATT agreements are far from perfect.

Nevertheless, production technology for agricultural goods in the USA, or increased subsidies for farmers in the European Community do affect the production by peasant farmers in India and elsewhere in the developing world through the workings of the world market. This is also true for industrial goods which substitute for agricultural products (e.g. animal feedstock made from fossil fuels) as well as for more or less identical products. Thus, prices in one part of the world affect what happens in the rest (and they also affect interfirm transfers by international firms). Even commodities which are conventionally considered to be non-tradeable, such as housing, may be prefabricated and internationally traded. Manufactured goods such as washing-machines are exported to developing countries where they substitute for the service of domestic servants. Major infrastructure projects are often carried out by overseas firms. The direction of change in the world economy suggests that this kind of exchange is increasing.

Even if we cannot agree that a 'perfect market' is a sufficient representation of reality, we might, at the very least, accept it to be part of the conservative vision of the future for a more preferable society, and so worthy of some critical evaluation. Many authors critical of the conservative approach question whether this concept of a market is useful, arguing that the 'trickle-down' theories of economic development lead to results which are too greatly contradicted by the experience of developing countries. Other authors, equally critical of the consequences of capitalist development, nevertheless stress the dominance of market forces. Murray (1971), for example, in considering changes in the organisation of transnational firms, argues that they likely to remain of limited significance when compared with the overarching discipline of the market economy. He describes transnationals quite simply as the 'bearers of market forces'. Our characterisation of world and domestic markets reflects this idea. For most of the experiments to be described later the idea of a perfect international market for commodities will be maintained. However, some key experiments will examine the implications of assuming rigid controls on the level of imports, in order to contrast this with the case of a perfect market. We are assuming, then, that an idealised market mechanism can capture some important elements of the real world – which does not necessarily mean accepting the conservative view of such a market optimising welfare.

We might ask, nevertheless, how great an impact the neglect of market imperfections is likely to have on experiments with the model. Several authors have examined this question — especially the effect of protective tarriffs on the performance of developing economies. Balassa (1981), a strong proponent of free trade, suggests that the cost of protection reaches 6 to 7 per cent of the GNP of several developing countries. In Brazil the figure is calculated to have been as high as 9.5 per cent (Balassa *et al.*, 1971). A calculation for Colombia by de Melo (1978), using a general equilibrium model (i.e. with feedback mechanisms similar to those we describe in this chapter), suggests a figure of 15.8 per cent. But it also argued that there can be important dynamic effects of import substitution and protection policies, principally to support the establishment of 'infant' industries in developing economies (or the restructuring of declining industries in richer countries). Here the debate is about how long, and what level of, protection is needed to develop the necessary technology, and to establish a niche in world markets. While our model does not address this question directly, our experiments exploring the consequences of autarkic policies, will illustrate the significance of import controls. (Although we do not make dynamic calculations with the model, these particular results are unlikely to be substantially different. A similar conclusion is reached by Cline (1982), with respect to his study of import-penetration.)

Given the assumption of a perfect market for goods, we might also assume a similar and partly independent market for international finance. In the social accounts described earlier, financial transfers, including development aid, capital flows and debt servicing, are combined, and balance the trade account. Estimates of levels of these transfers for each economy group for the years around 1975, are shown in Table 6.1. These data show a flow of aid and capital to the developing economies from the more industrialised economies, but also show substantial debt repayments. They do not reflect the more recent increase in development aid from the oil-producing economies to other developing regions. One major feature of international finance from industrial nations is the increasing importance of private (and usually non-concessionary) finance relative to that from governmental sources. World Bank (1981) data show that from 1970 to 1978 the non-concessionary finance increased fivefold to more than twice the level of concessionary flows of overseas development aid.

Table 6.1: *Aid, debt repayment and capital flows by economy group*

Economy group	1	2	3	4	5	6
Aid (1)	−8	−5	− 0.3	6	− 4	12
Capital flow (1)	−26	−11	0(2)	15	5	17
Debt repayments (3)	26	10	−14	−13	− 4	−5

(1) Data for 1975 is based on World Bank (1980), Table 10. Aid receipts are divided between Groups 5 and 6 in the ratios indicated by this table.
(2) Capital flows are taken to be the residual of other economies.
(3) The debt service is divided between Groups 1 and 2 in proportion to their total capital stock.

Despite the increasing importance of private sources of development finance, in our experiments we treat the total level of financial transfers as a variable which is determined in the context of particular development strategy. The apparent contradiction between taking the level of external finance to be an act of policy, and the level of trade for commodities to be the result of market forces, reflects the actual proposals of the international community. For example, at the Cancun Conference in Mexico in 1980, the United States administration, as the largest donor of aid, declared that aid would only be forthcoming to those nations which were prepared to accept free trade, both domestically and internationally.

Of course, trade is often limited by political or structural factors. As recent events have shown, pressure on developing countries to adopt more 'open' economic policies has been accompanied by a variety of restrictions on imports into the more industrialised economies. Any large increase in the net imports into any economy group is likely to result in corresponding changes in tariffs and quotas. Cline (1982), for example, takes a penetration level of 15 per cent of local production to be the trigger for the imposition of severe restrictions on imports of manufactures from developing countries. On this basis he argues that there is no possibility of all developing countries adopting an export-led strategy similar to that followed by the more successful newly-industrialised countries. While such an upper limit on trade is not included in our model, we follow a similar logic in interpreting its results.

Although many models treat the balance of payments deficit, and also development aid, as a result of the trade and savings gap (i.e. the imbalance between demand and supply), the scenarios suggest that the eventual size of the gap is equally a political decision: for example, IMF assistance to ailing economies may be usually conditional on restraints on domestic social expenditures or on other policy measures. The rescheduling of debt repayments to Western banks by Mexico, Brazil and Poland depend on political as well as economic consessions, and are determined at a governmental level as much as by private banks. (And Governments in industrial countries may subsequently arrange to redistribute the burden within their economy by increasing the national debt.) But in practice, foreign investment tends to respond to shifts in the relative rates of interest profitability between economies, and these particular phenomena are taken into account exogenously in our model experiments.

DOMESTIC MARKETS AND DISTRIBUTION

Commodity and financial markets within the groups of economies described by the economic model are also imperfect. These economic groups often comprise economies with competing interests. In the case of domestic economies, our model idealises markets for factors of production as well as those for goods. Factor prices (i.e. the wage rates for skilled and unskilled labour, and the rate of return on capital) are taken to be uniform across all sectors of each group of economies. Again, empirical research demonstrates that wage levels for people with equivalent skills or education vary widely, and that labour markets are often segmented and do not function well. On the other hand, it is not unusual for wage rates to be taken as an indication of skill level as well as of the marginal productivity of labour. (This procedure was adopted, for example, by Hufbauer, 1970.)

The assumption of uniform wage rates within economies, but not between them, implies that labour is perfectly mobile between the sectors of each economy, but cannot cross frontiers. International migration to the industrialised economies has been an important phenomenon in the past, and today immigrant workers from less-developed economies are an increasingly important source of labour in the oil-exporting Group 5 economies: repatriated wages are a major source of income for countries such as Egypt and Pakistan. Nevertheless, labour mobility across national boundaries is much more restricted both legislatively and socially than it is within boundaries. Because production techniques are relatively uniform within the economic groups, the approximation of a uniform wage rate for a given class of skills is more useful.

An equivalent approximation applies to the rate of return on capital. Across sectors within economies, a uniform rate of return is assumed. Further, capital is assumed to be mobile between sectors, implying that, within economies at least, investment takes place so as to bring the rate of return to a level at which no excess profits are gained. Again, this is clearly an approximation since profitability varies greatly even within sectors (where technologies of different vintage supplying products to different markets are employed).

Implicit here is the assumption that capital equipment and labour skills are interchangeable for the production of different goods. This is only partially true; but to the extent that agricultural land and labour can be used to grow luxury or basic crops, or manufacturing skills and equipment are switched between, say, prestige vehicles and public transport, the approximation is valid. In some of the experiments to be described later, we shall also consider the opposite assumption – that installed capital cannot be switched (the more conventional 'putty clay' approximation) and show that this is important to the overall level of growth in the world economy during periods of rapid restructuring.

Again, we may ask what the likely impact is on the model results of such assumptions. Calculations of the impact of imperfections in factor markets on the level of GNP, similar to those discussed above for international trade, are ambiguous. As an upper limit, de Melo (1977) calculated that the elimination of intersectoral wage differences might add from 5.7 to 13.3 per cent to the GNP of Columbia. (The higher figure arose from considering capital to be perfectly mobile between sectors, as in our model.) However, Johnson (1966) considered that neither the existence nor the elimination of factor price distortions is likely to be a strategic determinant of the level of national well-being.

The division of national income between capital and different kinds of labour, and the net level of transfers between households (as a result of redistributive taxation by government), is central to the discussion of income distribution. The procedure we use for distributing governmental income and expenditures between households, in order to estimate inter-household transfers, was described earlier. In the experiments, changes in the level of these transfers are treated in the same way as changes in the level of international transfers, as a deliberate result of state policy.

LABOUR MARKETS

When labour is especially abundant we should expect wage rates to be rather low, and to change little as the demand for labour increases. (In Lewis's (1954) famous model, labour in poor developing economies was assumed to be so abundant that wage rates were taken to be fixed.) When labour is in scarce supply wages are higher, and also more elastic to changes in demand. Working on the assumption that the smaller the reserve of a particular skill, the greater the bargaining power of that group is likely to be, we may relate the size of skill surplus to its share in national income. (Even in socialist economies, a variable classification of 'priority' labour exists with corresponding higher remuneration.) We have estimated the number of skilled people there are in the various types of economy relative to the demand for skills, using evidence based on UNESCO, ECE and ILO data. Despite the difficulties of definition this permits us to come to grips, to some extent, with the idea of under-utilised skills or under-employment, and gives an idea of the reserve of skilled and unskilled workers across each type of economy. For particular jobs there may be no suitably qualified person available, resulting in a skill shortage, however, across all occupations in an economy, there may well be a surplus of both skilled and unskilled people. It is this reserve which we attempt to measure.

Figure 6.1: *Relationship between wage rates and the utilisation of skills*

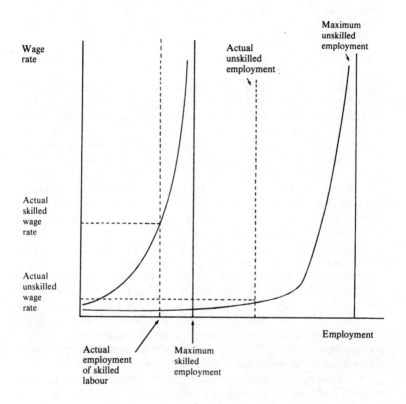

The estimated level of utilisation of skilled and unskilled labour (according to the definitions used for Table 5.9) are shown in Table 6.2. The data indicate that, in each region, unskilled labour is more abundant than skilled labour and, consequently, given the above argument, we should expect unskilled labour supply to be more elastic than that of skilled labour, as indicated in Figure 6.1. For Group 4 oil-exporting countries, the difference is especially marked. In the

Table 6.2: *Utilisation of skilled and unskilled labour by economic group*

Economy group	1	2	3	4	5	6
Skilled labour	77	87	85	69	95	83
Unskilled labour	68	64	77	42	⁻37	44
Ratio of skilled/ unskilled labour utilisation	1.13	1.36	1.1	1.6	2.6	1.9

developing countries in general, utilisation of unskilled labour is especially low, and only small shifts in wages occur with changes in the level of unskilled employment – they remain at subsistence levels, as in the Lewis model.

The degree of non-linearity in labour supply well above the subsistence wage will depend on, for example, the level of unutilised labour or the bargaining strength of low-income workers relative to other actors in the economy. The strength of this relationship in mixed and non-market economies (and between public and private sectors) is debated, but it provides a starting-point for a discussion of wage policy in terms of the negotiating strength or behaviour of different social actors. In particular, there is a distinction between types of labour and the supply curve. For India, Das Gupta (1976) points to a clear distinction between unionised labour for which the relationship may hold, non-unionised (casual) labour which is essentially tied to the subsistence wage, and high wage levels which are a legacy from the colonial era. He distinguishes India from some other developing countries where the highest salaries are fixed externally at the international level and a 'kind of oligopoly' is formed at the upper level which charges an extortionate price for its scarce services.

A decade ago, the general picture of labour supply in a developing economy was that of an abundance of unskilled labour and an acute shortage of skilled labour (e.g. Sutcliffe, 1971). Now, the situation, or at least our picture of it, has changed. On the whole, few skilled (i.e. qualified) people are actually unemployed, rather they occupy 'unskilled' jobs. Consequently, worldwide, the official level of unemployment among skilled people is relatively low (about 5 per cent) compared to that of unskilled people which may vary from 5 to 50 per cent (*ILO Yearbook*, 1978). Some authors (e.g. Acero, 1981) now report an abundance of skills worldwide (as in Table 6.2), but these skills are often poorly used, (e.g. qualified doctors in India working as taxi-drivers, economic graduates as bank clerks, etc.). This is not only a feature of developing countries; in many industrial countries the same syndrome is in evidence. In addition, there has been job simplification; jobs which used to involve a range of tasks, each requiring specialist knowledge or initiative, have been subdivided (e.g. the 'machining' operation separated from the 'tooling-up' operation). Thus some jobs have become de-skilled and repetitive. All these considerations mean we should be cautious in taking the empirical data in Tables 5.9 and 6.2 too literally, and be prepared in our later scenario building to take less restrictive assumptions.

CONSUMPTION AND INVESTMENT

The consumption behaviour of households was defined earlier in terms of the level of household expenditure (see Figure 5.8). The higher the level of *per capita* income, the greater the proportion of luxury goods in the consumption basket. As Table 6.3 shows, the

Table 6.3: *Regression coefficients for consumption and savings behaviour*

Dependent variable	Independent variables	Coefficient
Rate of investment	Constant	0.07 (0.018)
	Per capita disposable income	0.03 (0.004)
	Rate of profit	0.32 (0.04)
Correlation	R-squared	0.96
Proportion of basic consumption	Constant	−0.172 (0.029)
	Per capita expenditure	−0.269 (0.008)
Correlation	R-squared	0.99

Note: Figures in parentheses are standard errors.

correlation between the composition of consumption and *per capita* expenditures is high. For our experiments, therefore, it might appear appropriate simply to assume that, as the income of any particular household group rises, the consumption behaviour follows that suggested by the cross-sectional behaviour. This would imply that, as the income level of lower income households increases, or as the average level of income of developing economies rises towards that of the industrial countries, the same preference pattern will be observed. This is the underlying assumption in the conservative worldview. Cross-country studies (e.g. Lluch, Powell and Williams, 1977); Kravis *et al.*, 1982) show that this assumption is not without some foundation.

This view is criticised by the dependency and radical worldviews. If new styles of development, or improved levels of basic consumption of low-income groups, are to be achieved, then consumption patterns would have to deviate from this historical pattern, and the technology used to produce these goods would also have to be changed. Experiments in the next chapter examine the result of making explicit changes in the composition of consumption.

Similar considerations apply to the level of investment in each economy. The SAM estimates portray a high correlation between the ratio of total investment to national income, the *per capita* income of the region and the rate of profit. Other studies for developing countries (e.g. Fry, 1981; Summers 1981) also show a significant correlation between savings and the real interest rate. This relationship is given in Table 6.3. (With the simplified social accounts we are using, savings and investment are identical.) Again worldviews adopt different stances with respect to savings and investment behaviour. Consequently, although the relationship in Table 6.3 may be a useful guide for determining the level of investment, some experiments described in Chapter 7 will assume a national policy which departs from this.

TECHNICAL CHANGE

Finally, we describe the treatment of technical change. The principal assumption made for the experiments will be that the factor and intermediate input coefficients are fixed. Thus, there is no substitution between different skills, or capital and labour, or different types of intermediate good, unless an explicit change is made (e.g. to explore the introduction of unskilled labour-intensive production techniques in the least industrialised economies). In practice, there is some substitution between factors for individual production techniques. Economists have conventionally attempted to describe and estimate these relationships as Cobb–Douglas and constant elasticity of substitution production functions. But recent studies (e.g. Stewart and James, 1982, cited in Chapter 1), show that the technology adopted in the developing countries often does not correspond to the expected choices. The dependency worldview, in particular, argues that the reason for this is the technological domination of the industrial over the developing world.

The model does not incorporate explicit mechanisms to determine particular choices of technique. We shall, however, use it to examine the consequences of making explicit alternative choices. As indicated in Chapter 2, different worldviews assume quite contrasting relationships between production techniques and social organisation. (Technology is the combination of the two.) As a characterisation, more conservative theories assume, in effect, a form of technological Darwinism whereby techniques are developed in the quest for greater efficiency (profitability), and those which are unable to withstand the test of competition are discarded. In this worldview, the social context of innovation of new techniques is taken to have little bearing on the process of technological change.

The reformist perspective is more directly concerned with the impact of technological change on society, focusing on ecological as well as undesirable social impacts, but does not generally regard technology as a product of particular social relations. Given social objectives other than that of mere profit, more relevant choices of production tecnique can be made. More radical views assume that the production, choice and use of technology are governed by power relations (see Kaplinsky, 1983).

In the experiments, we shall make explicit changes in the technology described by the model. We use a Leontief-type production function with fixed coefficients, which may be modified to represent the introduction of a new technology. This assumption is also used for the intermediate input coefficients (described by the upper left-hand entries to the SAM shown in Table 5.1). Studies which attempt to explain changes in the input–output tables for different countries at a detailed level are very inconclusive. Data for the United States (Carter, 1970) and other countries (UNITAD, 1981) show that for industrial economies the coefficients are quite stable, and the few studies for developing countries (e.g. Gaiha's (1979) study for India) show rather few changes except an increase in inputs of energy to the agriculture sector. These studies suggest that, at the high level of aggregation in the model, it is reasonable to assume constant coefficients, and we take this assumption in our experiments, except when an explicit new direction for technical change is to be explored.

The behavioural relationships proposed for the model are chosen in order to permit exploration of a particular set of policies. These relationships are simple by any standards, and certainly belie the true

complexity of the world, as do the highly aggregated social accounts
to which they relate. An important point, therefore, needs to be made.
Given the over-simplified structure and relationships used, we might
expect the results of the model experiments to be exaggerated. Some
markets are assumed to operate perfectly (e.g. commodity trade),
while others (e.g. capital markets) function only in a restricted way.
Consequently, some of the balancing forces within economies are not
well represented.

The issue here is one of the timescale over which particular
behaviours operate. For example, local labour markets, like commodity markets, are assumed to clear rapidly (supply and demand
equalise) relative to other markets such as international labour
migration or major technical change. Even if we believe that the world
economy is never in equilibrium, and that new disturbances always
arise before the effect of earlier change has been accommodated, we
can still accept that the model displays possible, even probable,
tendencies. Especially if our concern, as in the experiments in
Chapter 7, is with the result of changes in the way markets operate
(e.g. restricting trade between industrialised and developing countries,
or changing the composition of consumption) then a comparison of
these tendencies may be indicative of the real world's response to
equivalent changes.

THE INTERPLAY OF STRUCTURE AND MARKETS

The results of our experiments depend critically upon *both* the
empirical structure of the Social Accounting Matrices, and the
behavioural assumptions described above.

To illustrate the interplay of behavioural and structural assumptions, let us briefly consider the results of some other economy-wide
models which explore the impact of fiscal transfers on redistribution.
these underlie conflicting responses to fundamental questions. Would
general equilibrium techniques have explored the impact of demand
redistribution, employment and growth. Notable amongst these are
the Korea model of Adelman and Robinson (1978), the Lysy and
Taylor (1979) model for Brazil, the ILO Bachue Philippines model
(Rodgers, Hopkins and Wery, 1979), and the Foxley's (1976) model
of Chile. These models are all multisectoral with multiple classes
described. Their results have one feature in common: whatever the
rules governing the division of surplus, if the impact of a transfer to
low-income groups is followed through a number of time-periods, any

improvement in distribution or gain in low-income employment tends to be systematically eliminated.

A number of explanations have been advanced for this. In particular, Lysy and Taylor (1979) argue, on the basis of a one-sector model, that the 'Keynesian closure' mechanism in their model, whereby prices are closely related to incomes and hence real wages are effectively pegged, is responsible. However, as Bigsten (1983) observes, Adelman and Robinson's model uses a neo-classical closure and the same phenomenon is observed. Thus, it is quite possible that the 'vanishing income' redistribution observed in the models may be a feature of the sectoral aggregation used in the models and their estimation, as much as the closure rules described. Indeed, in order to show the significance of this effect at a macro-level for Colombia, Thirsk (1980) assumes that the rich buy exclusively from large firms and the poor from small firms, and shows that the impact of fiscal redistribution to the poor is strongly augmented by this assumption.

The manner in which income redistribution may be dissipated because of dualism in the structure of production and consumption thus requires some examination in our model. For example, transfers to low-income groups may increase their consumption in the first round, but since this new consumption is produced using high-income labour and capital owned by high-income groups, most value added accrues to the latter groups. Alternatively, if the value added went to low-income groups, a multiplier effect based on successive increases in low-income consumption and production could be set up. Because of the way production and consumption data are independently collected and aggregated prior to their use in models, even models, such as ours, which do take account of structural heterogeneity may mask this effect.

We shall see that the distinction between low-income and high-income consumption goods has important consequences for the distributional effects of a modernisation strategy, such as that prescribed by the *status quo* worldview. One characteristic of this strategy, as it has proceeded in the past, is that as the techniques of production have become more sophisticated so, often, have the goods produced, with a change in the nature of the product from basic to luxury in the sense outlined earlier. Thus, for example, as production of shoes, soap, cereals or sugar shifts from traditional to more advanced techniques, the character of the production has often turned systematically from basic to luxury. Providing all incomes rise correspondingly, then the change may well be benefical to all income groups. But the switch to new products and new techniques of production through rapid modernisation, and an increase in the level of dualism in the economy, cannot provide an increase in the level of basic goods consumed by low-income groups, unless increases in their total wages offset the enforced changing composition of the consumption basket. (Changes in real wages depend, here, on the elasticity of labour supply characterised by Figure 6.1 and the change in factor inputs.) If the production techniques introduced displace unskilled labour so that their incomes fail to rise sufficiently, or even fall, the consumption of basic goods by low-income groups may decrease rather than increase.

Figure 6.2: *Modernisation with reduced basic consumption*

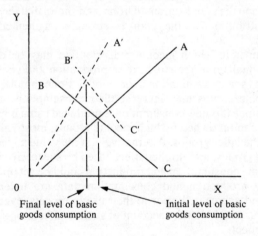

Figure 6.2 illustrates this point. The axes, OX and OY, are, respectively, the level of consumption of basic and luxury goods by low-income groups in a given economy. The ray OA shows the current composition of basic and luxury goods in the low-income basket and BC shows the current low-income budget. During the process of modernisation, the ratio of luxury to basic consumption increases — over time, OA is shifted to OA'. But if wages do not rise correspondingly, there is a relatively small shift in the low-income budget line from BC to B'C'. In this case, as shown in the figure, the new level of basic goods consumption by low-income groups will be *reduced*.

In practice, the actual impact of such a change would depend on the precise shift in the prevailing structure of the economy, the composition of production and techniques, and the dynamic framework within which the technology is introduced, including the determination of wage rates. Felix argues that, by contrast to the nineteenth-century growth of the now industrialised countries, in which

> dualism was moderate and self-liquidating . . . in twentieth-century LDCs, however, normal market forces tend to atrophy the artisan leg well before the modern sector can provide adequately offsetting employment. And . . . accelerating the growth of the modern leg often accelerated the atrophying, by hastening the premature displacement of artisan-built products with a cheaper or superior factory substitutes. (Felix, 1977, p. 204)

Of course, there need be no necessary association in the future between labour-displacing techniques and an increased luxury component in consumption goods. It is possible to cite contrary present-day examples here — luxury hotels, textiles, etc. may all be produced using the most labour-intensive artisan techniques and, conversely, the most basic products, e.g. plastic shoes, bio-gas plants and prefabricated dwellings are sometimes produced using the most automated. Despite this, the past experience of the developing economies indicates that, as long as both product and process innovation throughout the world economy are largely determined by the consumption demand of relatively high-income groups and factor costs in the already industrialised economies, this trend will tend to be the rule.

Authors such as Felix (1977) and Stewart (1977) emphasise the impact of these processes on developing countries. But dualism also exists in the most industrialised economies (e.g. Averitt, 1970). Thus, while differences across income groups are less in industrialised economies, the process whereby low-income groups may be systematically obliged to cut their basic consumption because of changes in the production system is also seen there. For example, a shift to private vehicle ownership, luxury packaging, or out-of-town shopping centres has often been followed by reductions in the quality of public services and other amenities to the poor and immobile (see, e.g. Hillman and Walley, 1980). One consideration our scenarios raise for the future is whether a new form of dualism could arise in the less dynamic industrial economies with the rapid introduction of very labour-displacing production techniques (e.g. microprocessor-related technologies) worldwide.

In both industrial and developing economies the relative neglect of the welfare of the lowest income groups relates to their overall lack of effective demand. In developing economies with a high level of maldistribution and absolute poverty, low-income groups may be a vast majority, but their total demand is relatively small. In industrial countries the lowest income groups are relatively better-off, and fewer will be perversely affected overall by modernisation.

The alternative 'human needs'-oriented strategy, explored in Chapter 7 for the least industrialised economies, aims to create a situation whereby low-income labour producing basic goods for their own consumption would be the principal dynamic of a redistributive growth strategy. This would imply an alternative technological route with a relatively higher production of basic goods via techniques employing somewhat less labour-displacing techniques.

The possibilities for increasing levels of consumption of basic goods through such a strategy are indicated in Figure 6.3. In this case, compared with Figure 6.2, the ray OA showing the composition of production shifts to OA" and the low-skilled group's budget line BC shifts to B"C". As shown in the figure, even if the production techniques assumed here did not give rise to an overall increase in wage income, it is still possible for the level of consumption of basic goods to increase. This suggests that the production system of the traditional sector could provide the starting-point for the alternative strategy.

Figure 6.3: *Alternative strategy to increase basic consumption*

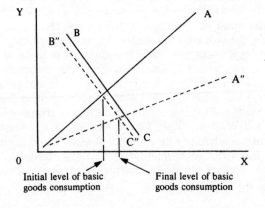

Although these two illustrations do not, by themselves, explain the results of experiments with the model given in Chapter 7, they do provide insights as to why results of the kind we describe are obtained when distinctive economic structures and market mechanisms are represented in a global model.

Some of the results of the model in Chapter 7 may seem surprising at first sight, until we remember the paradoxes introduced when simple models are extended to include many agents and diverse behaviours. The model treats the world economy as an *integrated entity*, and so, unlike single-sector or partial equilibrium models, redistributes production and consumption worldwide in what may be a complex manner. Although the results are relatively sensitive to the precise changes introduced, the explanation of their characteristic pattern is often simple enough. Factor prices are determined locally, and commodity prices determined internationally. Technological change in one sector of one economy requires that factor prices are adjusted first across that economy; but, in contrast, commodity prices are adjusted across the entire world. Thus, the shift in commodity prices results from a change in the *average* production technique across the world. Changes in technology, therefore, lead to modest shifts in local factor prices, but only to comparatively small changes in world commodity prices. These differential factor and commodity price changes, then, together dictate the subsequent behaviour of the model — in particular, changes in the composition of production and the domestic and international distribution of income.

7 Development versus the Market: Experiments with the Model

INTRODUCTION

In this chapter, we use our model to explore major policy instruments of the development strategies described in Chapters 2–4. Each worldview lays emphasis on particular policies: the reformist view, for example, stresses the importance of international and domestic income transfers: the conservative view, the importance of modernisation of the production system and maintenance of free trade: and the radical perspective emphasises the need to break away from the world market economy. Our experiments will attempt to evaluate these proposals from the standpoint of our model assumptions and structure. Because the explanation of transfer processes is usually less complicated than that for technology or trade policy, we shall begin by examining the impact of international development aid. Generally, we shall be changing only one or two parameters in the model: thus the experiments test how sensitive major variables, like income distribution, are to the policy instruments included in the model. In Chapter 8, we shall incorporate the findings of the individual experiments into a more integrated vision of how the world economy may develop over the next forty years.

DEVELOPMENT FINANCE AND THE MARKET

Development finance is a key policy instrument of reformist strategy. The first part of this chapter describes a small number of experiments which simulate the transfer of a fixed amount of development finance from the most industrialised to the least industrialised economies, under different policy conditions. The amount involved is in line with the suggestions of the Brandt Commission (1980) and other studies.

The Commission reported that the aggregate needs of all the least

170

developed countries for external capital, to support a 5–6 per cent growth rate, is estimated by UNCTAD as US$11 billion per year during the 1980s and US$21 billion per year in the 1990s. Steinberg and Reger (1978) project developing countries' total requirements of a yearly average of US$50–3 billion (in US$ 1974) over the period 1976–80. The Brandt Commission suggests that, since about 40 per cent of development aid goes to countries with *per capita* annual incomes above US$400, around US$30 billion would be needed as development assistance for capital projects to the least developed group of economies. The amount required appears to be steadily increasing. In his report to the 6th UNCTAD Conference in 1983, the Secretary-General estimated that the extreme finance needs of developing countries in 1984–5 could be as much as US$ 90 billion.

The income of the least developed economies is about 4.1 per cent of total world income and their population is 61 per cent of total world population. For the purpose of the experiments, it is assumed that development finance to developing countries is transferred with the aim of ultimately benefiting their lower income households, as would be suggested by reformists. Thus, in some of the following experiments, we assume that investment income arising from the transfer accrues to this poorest group, in addition to other income arising from changes in their level of employment. It is also assumed that the full amount of the transfer is from the high-income group in the most industrialised economies.

There are two main steps to be explained. The first is the initial transfer and the purchase of new capital goods; the second is the impact of this new production capacity on the domestic and international economy.

Step 1: making the investment transfer
In the first experiment, a transfer of US$30 billion is made: it is subtracted from the investment income of the high-income group in the richest economies, and added to that of the low-income group of the poorest economies. Balance of payments levels are adjusted accordingly. The investment goods are purchased but not installed, so at the first stage of the transfer no new employment or investment income is generated. Neither is any interest or repayment due, so it is (for the moment) immaterial whether the transfer takes the form of a gift, loan or direct investment.

Since transfer and investment changes balance, there is no change in the levels of consumption of either household group or in their combined demand for investment goods. The investment goods which would previously have been purchased by the richest group and installed in their domestic economies are simply transferred to the poorest. In effect, therefore, the development assistance has the appearance of 'tied' aid (i.e. the recipient has purchased goods from the donor country, even though there may be nothing in the terms of the aid agreement to ensure the purchase from the donor). (Generally, when aid is tied, its value is reduced; Singer and Ansari (1978), for example, suggest that tied aid has a purchasing value about 80 per cent of untied aid. In the experiment, however, aid is assumed to maintain its full purchasing value.)

Table 7.1: *Summary social accounts for all economies: initial transfer of investment aid*

Economy group	1	2	3	4	5	6
Output:						
Luxury	2756.4	1088.2	937.1	276.6	124.3	95.5
Basic	1049.1	509.2	783.2	422.1	172.5	431.5
Investment	919	431.3	423.7	169.9	83.9	97.6
Factors:						
Skilled	586.4	339.3	137	86.8	60.2	42.5
Unskilled	1138.3	488.2	389.9	274.9	102.6	150.6
Capital	552.3	290.5	560.1	180.3	81.2	90.9
Households:						
High-income	884.2	475.3	469.6	216.6	66.5	85.5
Low-income	1279.6	650.4	634.1	358.4	178.5	253.3
Net trade:						
Luxury	27.1	–11.8	–14.9	–16.3	27	–11
Basic	–24.9	–8.9	32.1	8.5	–13	6.1
Investment	111	13.1	–34	–25.2	–15	–50

Notes: World price of commodities: Luxury = 1.0000 Basic = 1.0000. Amounts in US$ bn (1975): prices are relative to investment goods.

The result of the experiment is shown in Table 7.1. A comparison with Table 5.16 for the base year suggests that rather little has happened! Only levels of trade have changed. Nevertheless, many important mechanisms have operated to bring about this result and satisfy the condition that commodity prices worldwide will be equal in equilibrium. The principle mechanisms which lead to this result are shown in Figure 7.1. To understand the result we should consider what would have happened had the purchase of the additional investment goods been made in the recipient economy or a third-party economy: this would simply have had the effect of increasing the price of investment goods in that economy above the world price. Additional labour would have been required for their production; given the positive relation assumed between labour supply and wages shown in

Figure 7.1: *Transfer of investment aid to the least industrialised economies*

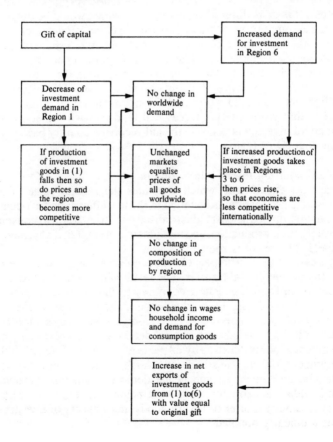

Figure 6.1, this would increase the price of labour in that economy (and hence of all goods produced). Correspondingly, the price of goods in the donor economy would fall. Even if some of the investment goods were to be purchased other than in the donor economy, this would be compensated for by shifts in production for existing demand. Price differences could only be supported by changes in the balance of payments in the model, or by other means not included in the model (and not consistent with the operation of a free market). But, in this transaction at least, the model suggests that although the donor country does not gain, neither does it lose.

While this may appear to be something of a non-result, it is a general equilibrium solution derived as a consequence of the assumptions of the model as all factor and commodity markets adjust together. It is quite different from the result which would be obtained with the assumption more commonly used in global models (see Cole, 1977), that each nation provides a more or less fixed share of world exports of each commodity. This last assumption, in effect, ignores differential changes in costs of production such as those described above, and so indicates that only a part of the US$30 billion development aid would return to the donor economy. But our model suggests that world market forces will tend to push the additional demand for investment goods in the aid-receiving country back to the donor country or region.

Step 2: installing the new investment
In the next experiment, which simulates the second stage of the transfer, the new capital is installed in the least developed economy. The population in the least industrialised countries starts to consume the fruits of new production, and repayments on capital, if any, have to be made. The result is shown in Table 7.2.

It is evident that the impact of the installed capital is quite different from that of the intitial transfer. The result is not simply a switch of consumption from the richest countries to the poorest. Significant shifts in the structure of production and trade occur worldwide. Luxury production in the recipient economy increases by some 90 per cent, while basic goods production declines by 13 per cent, and investment goods production by 22 per cent. A relatively small shift in employment and incomes belies the larger shift in production and trade. Because incomes, and hence domestic consumption levels, change little, the economy shifts from being a net exporter of basic goods to a net exporter of luxury goods; net imports of investment goods are nearly doubled.

Table 7.2: *Summary social accounts for all economies: installation of new capital*

Economy group	1	2	3	4	5	6
Output:						
Luxury	2689.5	1081.5	930.5	274.9	123.9	181.4
Basic	1106.2	513.1	785.3	421.6	172.7	383.4
Investment	930.9	435.3	428.3	172.2	84	80.2
Factors:						
Skilled	585.7	339.2	137.4	86.6	60.2	42.5
Unskilled	1141.5	489	389.2	274.5	102.5	150.7
Capital	550.6	290	560.3	180.6	81.3	94.2
Households:						
High-income	912.2	474.9	470	216.6	66.5	85.5
Low-income	1282.4	651.1	633.5	358.2	178.4	226.8
Net trade:						
Luxury	−40.4	−18.7	−21.6	−18	26.7	72.1
Basic	31.2	−5.3	34.3	8	−12.7	−55.5
Investment	92.3	16.3	−29.4	−23	−15	−41.4

Notes: World price of commodities: Luxury = 0.99991 Basic = 0.99994 Amounts in US$ bn (1975): Prices are relative to investment goods.

By contrast with the first stage of the transfer, the donor country alone does not provide these additional imported investment goods to the recipient economy. These exports are now provided also by other countries, notably the newly-industrialied group, and also the Groups 2 and 3 industrial economies. Group 1 countries provide, especially, the additional demand for basic goods. What this result suggests, for example, is that in an unfettered world economy, if the USA (in Group 1) provides aid to Pakistan (in Group 6), the latter is liable to produce such things as export-quality shoes and run down its agriculture, and then import wheat from the United States and purchase milling machines from South Korea (in Group 4).

By comparison with the shifts of production and trade, there are rather small changes in income. At first guess, we might have expected that the US$30 billion new investment would have raised the total income of the economy by an amount more or less *pro rata* to the existing stock (about US$790 billion). This does not happen, because market adjustments leave employment and wage levels relatively unchanged in the new equilibrium. The income of the recipient group in the least developed economies only increases by an amount roughly equal to the increase on investment income (US$30 billion $\times$ the rate of profit). The income of the high-skilled group in these economies is practically unchanged.

The Group 1 donor economy, by contrast, displays more marked shifts in employment. The number of unskilled jobs increases; that of skilled jobs decreases. Consequently, the income of the low-income group rises in the most industrialised countries, while that of the high-income group falls. This is despite the fact that the direct transfers in this experiment are identical to those in the base year. The income distribution in other economies also changes, becoming more equal in less dynamic industrial economies and less so in the newly-industrialised and centrally-planned economies.

In obtaining this equilibium a rather complex process of adjustment has taken place. The reasons for the changes will be considered for the recipient economies, the donor economies and the rest of the world in turn. Often, authors of general equilibrium models avoid attributing any economic significance to the *process* of adjustment, preferring rather to demonstrate that the outcome has an internally consistent rationale (see for example De Melo, Dervis and Robinson, 1982). The problem in providing a causal explanation is that, since every variable is mutually dependent on all others, and since there are no time lags in the model, the result of the experiment should strictly be thought of as the full working-through of all market processes, rather than as a sequence of events. But when working in a wider context, of the sort provided by chapters 4 to 6, it seems quite appropriate to attribute causal significance to the model results.

Processes in the recipient economies
The underlying assumptions of the general equilibrium model and the precise magnitudes of the factor supply and technical coefficients shown in Tables 7.2 and 7.3, are central to the shifts observed in the experiment. When additional capital is introduced into an economy, additional production becomes possible. In most cases, this will require that additional labour is employed, which causes the price of labour to rise (in accord with the labour supply equation). Capital will then be relatively cheaper, and production in the more capital-intensive sector will become more favourable than hitherto.

In the present experiment, in equilibrium, the employment levels of skilled and unskilled labour in the least developed Group 6 economies rise slightly. Capital becomes relatively and absolutely cheaper and the rate of profit falls. Because the luxury goods sector in this economy is the most capital-intensive, its competitiveness in world markets for goods and domestic markets for factors of production is

Table 7.3: *Summary social accounts for all economies: new investment in the most industrialised economies*

Economy group	1	2	3	4	5	6
Output:						
Luxury	2789.3	1077.8	926.4	271.9	125.8	94.3
Basic	1068.6	509.6	773.7	416.1	170.3	432.3
Investment	871.5	446.5	445	180.7	84.9	98
Factors:						
Skilled	589.6	340.6	139.1	86.4	60.4	42.6
Unskilled	1144.7	490.6	388.7	273	102.2	150.7
Capital	547.9	287.8	559.2	181.6	81.2	90.7
Households:						
High-income	914.4	474.7	470.9	217	66.7	85.5
Low-income	1284.5	652	632.8	357	178.1	223.3
Net trade:						
Luxury	55.4	–24.1	–26.5	–21.1	28.4	–12.2
Basic	–7.3	–9.7	22.7	2.7	–15.3	6.9
Investment	35.1	26.1	–12.9	–14.6	–14.1	–19.5

Notes: World price of commodities: Luxury = 0.99968 Basic = 0.99968. Amounts in US$ bn (1975): Prices are relative to investment goods.

strengthened. This is not an automatic process. The model functions so as to redistribute capital and labour in each economy to achieve uniform factor prices across the economy. This implicit restructuring of the economy makes an important contribution to the results.

In this experiment, the capital used to produce different commodities is interchangeable, although the factor proportions in each sector remain constant. Capital is neither written off, nor taken out of production during the adjustment. For example, if production of luxury motor cars rather than trucks becomes more favourable, then the output of the plant would be switched from trucks to private cars. Similarly, land used to produce staples such as maize is switched to production of hard wheat for export. There are obviously limits to such malleability in the use of capital and, in some experiments, this assumed homogeneity of capital may well exaggerate the likely effect of a change in model parameters. (The implications of a loss of production capacity during the restructuring process is considered later.)

Figure 7.2: *New production capacity in the least industrialised economies*

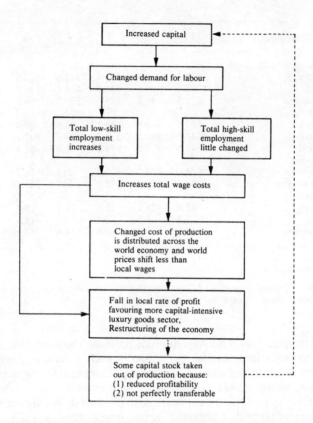

The processes which lie behind the changes observed for the least developed Group 6 economies are shown in Figure 7.2. As noted above, the decline in the relative price of capital in the least developed economies favours the production of luxury goods. We might expect that the small increase in the price of skilled and unskilled labour would also promote production of basic and investment goods, since the capital and basic goods sectors in this economy are the least intensive in their use of unskilled and skilled labour, respectively. However, these tendencies are offset by the larger fall in the price of capital.

This result has direct implications for the reformist strategy for global development described in Chapter 3. The Brandt Commission, in particular, suggest the development finance should be directed specifically to production of food (which makes up most basic production in the poorest economies). The results of the model indicate that if the new capital was restricted to the production of basic goods, then the rate of profit or wages in this sector would have to be lower than in the rest of the economy. In this situation, entrepreneurs will invest in more profitable activities or labour move to better-paid jobs. Consequently, the initial attempts to tie down the development investment to the basic goods sector would soon be frustrated by the migration of private capital out of the sector.

The donor economies
The increased competitiveness and increased production of luxury goods in the least developed economy is matched by a decline in their production in all other economies. This releases production capacity for use in other sectors. Which sectors gain depends on their relative factor inputs and the supply elasticities of factors.

In Group 1 economies, in particular, the level of employment, and hence of wage rates for low-skilled labour, increases. The decrease of employment levels and wage rates of skilled labour favours the expanded production of luxury goods, and is least favourable to that of basic goods. Despite the fact that the changes in wage rates are not conducive to increased production of basic goods, this sector expands. This is because of the capital good sector in this economy has especially low capital intensity. The fall in the price of capital is thus relatively more favourable to basic goods production, this more than offsets the effect of adverse wage shifts on the basic goods sector.

Processes in other economies
The changes observed in other economies may be explained in similar terms. In the newly-industrialised economies, for example, the level of employment for both categories of labour falls while the price of capital rises. The fall in the price of unskilled labour favours the production of basic goods. However, the fall in the price of skilled labour favours the production of investment goods more than production of basic goods; the increase in the price of capital also encourages the production of investment goods. Consequently, production of investment goods increases, and this economy increases its exports of investment goods to the least industrialised economy.

The net outcome of any experiment depends on the relative factor inputs across economies as well as across sectors within economies. This means that in some cases the explanation of the observed effects can be even more complex than those described here! Fortunately, the shifts in commodity prices observed in the model are relatively small when no changes in production techniques are assumed. For these experiments, therefore, changes in competitiveness arising from the cost of intermediate inputs to production, need not be considered.

Alternatives to development aid

Direct foreign investment and investment loans. The previous two experiments assumed that the transfer of all benefits of the development assistance would be to the poorest and politically weakest economic group. But this has a ring of utopianism. Even the Brandt Commission emphasised the self-interest of the North in providing finance to the South. It is sensible to look at other kinds of investment to explore how they affect the income of different social groups, especially since they are recommended by many conservatives.

One way for foreign capital to reach the developing countries is as direct investment. In this case, rather than donate aid, the high-income group in the most industrialised economies invests directly in the Group 6 economies, and the profit income from the new stock is repatriated directly to the investors in the Group 1 economies. The income of this group, therefore, rises by some US$5.5 billion (US$30 billion $\times$ the rate of profit in the Group 6 economies). Conversely, compared to the previous experiment, the income consumption expenditure of the low-income group in the least developed economies falls by approximately the same amount. The resulting excess production from the least developed economy is now largely exported to the donor economy and paid for out of its additional profit income.

Alternatively, we can treat the transfer as a loan. In this case, the repatriated income to the Group 1 economies is reduced to US$2.3 billions (US$30 $\times$ the rate of profit in Group 1 economies). The major conclusions are, however, little changed. With either direct investment or loans, the structure of world production is little altered from that displayed in the previous experiment, when the finance is provided as a gift (Table-7.2). The shift in the production structure is again large, and would, in practice, represent substantial disruption of the economy. For example, large-scale migration from rural to urban areas is implied, which could cause the economy to become significantly less self-sufficient in the production of staple commodities and capital goods. Thus, the economic advantages from this kind of direct investment to the recipient economy would appear to be somewhat muted since *per capita* income gains are small while the adverse social implications may considerable.

In the light of these results, we should also ask whether the income of rich households in the Group 1 countries would be greater if they invested in their countries, and what the impact of this would be on the poorest countries. Although this is certainly not an act of development aid, the result of this experiment given in Table 7.3 provides an instructive contrast to our previous findings. Adding US$30 billion capital to the existing stock in the most industrialised economies obviously must raise the total income of capital owners. But this increase is not as large as when the same investment is made on their behalf in a developing economy with a significantly higher rate of profit.

Figure 7.3: *New production capacity in the most industrialised economies*

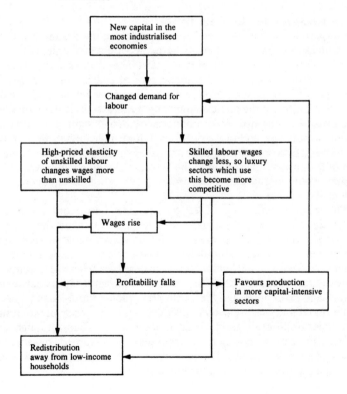

Given the international differences in the rate of profit, and the assumption that all the net surplus is repatriated, this result is not surprising. The impact on income distribution, however, is less straightforward. The new investment in the industrial countries leads to redistribution of income away from low-income households there and they are now *worse-off* than if the new investment has been placed in the least industrialised economies. This result arises from the high elasticities of supply for unskilled labour in the most industrialised economies, and the relatively low capital-intensity of the investment goods sector there. The processes underlying this result are shown in Figure 7.3. This result appears especially significant in view of the lukewarm or hostile attitudes of many trade unions in the North to investment transfers or development assistance to the South. It suggests that unskilled workers in the North could actually gain from the international restructuring of production stimulated by development aid and foreign investment.

Conclusions on development finance

The experiments just described show that, given the assumptions in the model, transfers of capital — whether as direct investment or as a gift in isolation from other forms of assistance — have rather small medium-term impacts on the income of the least industrialised economies. This is not to deny that there may be an important cumulative effect if the additional income generated is systematically reinvested. For example, if the finance is treated as a gift to the lowest income group, the additional US$3.6 billion investment income is roughly equal to one-third of their annual savings, and is about 10 per cent of the annual investment in the least developed economies. This amount would increase the total capital stock by about 2 per cent per annum.

The main beneficiaries of the transfer are the donors, although there are complex redistributional effects throughout the world economy. One result repeated in all the experiments is that the transferred investment stimulates redistribution of income in the donor economy, when it leads to a shift in production between sectors. These shifts arise because of the variations in factor inputs between sectors and economies and the variation in factor supply elasticities. Different techniques are used to produce the same goods in different economies or different goods in any one economy. Thus, although no changes in sectoral production techniques are assumed, *aggregate* factor proportions across the economy will shift, and so change the total demand for skilled and unskilled labour.

The extent of these shifts depends on the heterogeneity in the production structure of each economy, which tends to be greater in the less industrialised economies. Nevertheless, as the experiment given in Table 7.3 shows, the redistributive effects of new investment in the industrial economies leads to a divergence between national interests. The impact of changes in overall levels of employment on relative wage rates and hence the rate of profit, depends also on relative wage rates and hence the rate of profit, depends also on the relative supply elasticities for skilled and unskilled labour, which again very markedly across the economies considered. These relationships become even more important when we consider explicit changes in production techniques, as in the next section of this chapter.

In all the experiments, the ratio of imports to consumption increases for the least developed economy; in other words, a higher proportion of consumption is based on imported produce. The level of imports of capital goods increases in each case, although it is significant that it is the newly-industrialised economies, rather than the most industrialised economies, which provide these additional imports.

One arguable limitation of the present model in evaluating reformist proposals is that it treats the level of stock in each economy as fixed, and assumes full utilisation of capital stock. The Brandt Commission, for example, emphasises the IMF view that 'massive' transfers are particularly important to the industrialised economies when, as in the present recession, there is under-utilised capacity. It can be argued, however, that to bring unutilised capacity back into production in a market economy would require an increase in the rate of profit — in the same way that the model assumes that an increase in the wage rate is a prerequisite for an increase in the labourforce. In all the experiments reported so far, the rate of profit in the most industrialised economies is rather stable, shifting by less than 1 per cent of its rate in the base period. Consequently, unless capital supply was very elastic, it is unlikely that modifying the model in the way suggested would greatly affect the results. (This topic is considered later in the context of the restructuring of the economies.)

The main conclusion that we draw from these experiments is that domestic and international market adjustment can considerably reduce the beneficial effects of development finance. This result repeatedly emerges, under a set of plausible assumptions, and it is likely to do so with quite substantial modification of the assumptions. Whatever the intrinsic merits of the reformist proposals (e.g. massive development assistance to Third World countries), conditions such as those laid out by the present United States administration, that recipients of such assistance should operate as market economies, may well nullify any potential gains and do nothing to improve social distribution. Furthermore, the effect of development finance appears to weaken those basic production sectors (such as agriculture) which the Brandt Commission hopes to strengthen: in all the experiments, luxury goods production gains at the expense of basic goods production. Thus countries which adopt open-door policies towards foreign investment and goods in order to obtain aid may experience a decline of the rural sector, increased migration to urban areas and a worsening of income distribution. This may apply, for example, to countries like Chile, which received substantial aid from the USA once it turned towards free-market economic policies.

Finally, we should note that while these results may appear counter-intuitive or contentious, they are not unparalleled in the literature. We mentioned in Chapter 6 the theoretical studies of Gale (1974) and Chichilnisky (1980). The latter author takes a two-commodity, two-economy, three-income group model (a simplified version of the model we use) to explore the possible effects of aid on welfare. These studies identify some of the conditions under which international markets can thwart development efforts. We have here shown that when this theory is reproduced in an empirical computer model, with parameters based on real world data, the effects of aid do indeed appear to be perverse. Our conclusions parallel these reported by more astute observers of the impact of food aid (e.g. Singer and Anasri, 1978).

CHOICE OF TECHNOLOGY AND THE MARKET

The experiments described in the previous section show the results of increasing investment assuming that there were no changes in production techniques. By contrast, the experiments which follow compare the implications of adding distinctly different technologies to the existing production system. These technologies are characterised, following the emphases of different literatures, as 'appropriate', 'modern sector' and 'microprocessor-related'. Each of these technologies has distinctive characteristics with respect to the factor inputs to production, the type of product produced, and the availability of the technology itself.

The implementation of microprocessor-related technology is an important feature of the conservative strategy, which expects these technologies to cut production costs dramatically. Some studies (e.g.

Kaplinsky, 1980; Rada, 1980) point to the challenges that the adoption of the new technologies based on microprocessor devices and associated information systems by the already industrialised countries poses to developing countries. Industries such as textiles and other manufacturing, for example, which have in the last decades migrated to developing countries, may now move back to the more industrial economies.

The new technologies, in addition to employing a different composition of labour and capital from existing techniques, also employ a different composition of skills. Thus, the introduction of new technologies in a given sector is likely to affect relative levels of employment and the distribution of incomes among different categories of labour within economies, as well as that between aggregate labour and capital. Further, changes in one sector of one economy will be communicated to the rest of the world through domestic and international markets. Consequently, changes in technology will have both direct and indirect effects on the worldwide distribution of income.

Some authors (e.g. Sarabhai, 1968; Clark, 1981) have suggested that developing countries should not adopt intermediate or appropriate production technologies and their products. The least developed countries, not yet locked into the already obsolete modern-sector technologies of the North, should, instead, move directly to adopt the more radical new technologies wherever possible.

In contrast to new technology, appropriate technology is usually defined as having relatively low requirements for capital, to be more labour-intensive overall (yet less demanding of formally trained labour), to produce less standardised products, and so on. Appropriate techniques should, therefore, increase demand for the abundant factor of production (unskilled labour) and reduce demand for expensive capital and skills in the least developed economies. Arguements for its adoption (e.g. Singer, 1977; Diwan and Livingstone, 1979) stress the greater employment-creating potential of the technology by comparison with modern-sector technologies.

We now examine each of these options for technology policy in turn. In these experiments, it will be assumed that US$30 billion of new capital will be devoted to one or other of these technologies. The purpose is to compare the income-creating and distributional impact of the various technologies, and, in particular, to explore to what extent a policy to invest in such techniques is affected by domestic and international markets.

NEW TECHNOLOGY IN THE MOST INDUSTRIAL COUNTRIES

We first explore the possibility that the most industrialised developed economies of Group 1 could, in fact, successfully introduce micro-processor-related techniques into luxury goods production. (This would certainly be in line with the conservative prescription set out in Chapter 4.) Any estimate of the changes in labour productivity on a sector or economy-wide basis as a result of the introduction of new technologies is bound to be controversial. On the basis of Bessant's (1980) synthesis of many studies of the impact of microprocessor-related techniques on employment in twenty-nine sectors of the economy, we assume that the aggregate unskilled labour-output ratio of microprocessor-related techniques is half that of current modern-sector techniques (i.e. the same level of production may be achieved with only half the unskilled labour requirements imposed by current technologies). Changes in capital requirements are even less certain; again on the strength of Bessant's study, these are taken here to be the

Table 7.4: *Summary social accounts for all economies: new technology in the most industrialised economies*

Economy group	1	2	3	4	5	6
Output:						
Luxury	2961.3	1003	855.4	260	115.2	88.8
Basic	907.1	572.6	841.2	428.5	181	436.9
Investment	861.8	458.8	447.7	180.5	81.9	99.1
Factors:						
Skilled	593.4	335.5	138	85.1	58.9	42.9
Unskilled	1134.6	496.1	381.8	273.8	102.7	151.8
Capital	553	288.2	564.8	182	82.2	89.9
Households:						
High-income	922.3	469.8	474.2	215.9	65.9	85.2
Low-income	1275.6	657.6	627.1	357.9	178.9	224.1
Net trade:						
Luxury	224.8	−95.1	−97.8	−32.7	18.7	−17.8
Basic	−167.1	52.8	91.6	14.6	−3.1	11.2
Investment	25.6	34.5	−10.5	−14.8	−16.5	−18.2

Notes: World price of commodities: Luxury = 0.99946, Basic = 0.9998. Amounts in US$ bn (1975): Prices are relative to investment goods.

same as those current in the luxury sector in the Group 1 economies.

Table 7.4 shows the results of assuming that US$30 billion of new investment goes into the installation of these new technologies in the luxury goods sector in the most industrialised Group 1 economies. The change in technology increases the competitiveness of luxury good production in the countries relative to their other production, and relative to luxury goods production in other economies. Not surprisingly, the results show a large increase in the production of luxury goods and a decline in basic and capital goods production in Group 1, with corresponding changes in the structure of production throughout the world economy. The GDP of Group 1 countries rises as a result of the new capital. This also leads to an increase in the demand for skilled labour, but there is a fall in the demand for unskilled labour. The increased competitiveness in the economy has, therefore, been insufficient to compensate fully for the labour-displacing effects of the new technology. Furthermore, the fall in prices brought about by the labour-saving production techniques is insufficient to raise their real incomes (i.e. the amount of goods actually purchased) above the base year level. By contrast, the wages received by high-income households increase, as does their invest-ment income. Thus, overall, there is an increase in the GDP of the countries adopting the new technology, but a worsening of income distribution.

The results for other economies are also mixed, and some economies and groups benefit indirectly from the change. In particular, the employment of low-income groups in the least industrialised economies increases (due to the growth in basic goods production for export to the most industrialised economies).

Restructuring and decline
The restructuring of the world economy suggested by the last experiment is considerable. In practice, economies would adjust in several ways. The less competitive economies might, for example, devalue their domestic currency to counter their loss, in part, and to spread out the fall in income across all factors of production. (In the model we represent a shift in the exchange rate as a uniform change in all local factor prices and transfers between households, while local commodity prices are held equal to the world price.)

A second adjustment arises because, in practice, capital in production cannot be switched between sectors as easily as the model implies. Some capital would be prematurely written off in the process of restructuring. Consequently, the preceding experiment overestimates the new volume of output in each economy, and hence in the world as a whole. To take account of this, we now replace the 'putty capital' assumption (that treats installed capital as having complete mobility) by a 'putty clay' assumption. With this, only part of the total capital (some 85 per cent) can be switched to production of other goods.

Table 7.5: *Summary social accounts for all economies: New technology in the most industrialised economies (with scrapping)*

Economy group	1	2	3	4	5	6
Output:						
Luxury	2836	1001.6	939.7	276	120	87
Basic	972.7	543	777.2	445.7	179.7	437.7
Investment	904.2	489.5	414.4	144	79.1	99.1
Factors:						
Skilled	583.7	332	133.3	85.9	58.5	42.6
Unskilled	1124.2	491.5	383.5	278.1	103.5	151.7
Capital	563.9	292.2	564.3	178.7	81.9	90.1
Households:						
High-income	921.1	469.4	469.2	214.6	65.3	85.1
Low-income	1267.5	654.1	628.7	361.1	179.6	224.1
Net trade:						
Luxury	112.2	-94.6	-6.1	-15.3	23.3	-19.4
Basic	-96.5	24.5	31.7	32.6	-5	12.7
Investment	67.5	62.4	-42.3	-50.3	-19.4	-18.1

Notes: World price of commodities: Luxury = 1.00013. Basic = 1.00058. Amounts in US$ bn (1975): Prices are relative to investment goods.

The consequences of this change are shown in Table 7.5. In this new experiment, the GDP of the Group 1 industrialised countries now falls, as does that of most other economic groups. With lower production in the rich countries, the employment of both skilled and unskilled workers falls. The situation of low-income households has further worsened, and the total income of high-income households has fallen below the base level, despite an increase in the rate of profit as existing capital is scrapped. Again, the implications for other economies are varied, but the income of the poorest countries is increased above the base year level, as is that of the newly-industrialised group.

These results are sensitive to the assumptions about the new technology. For example, if we assume that the new technology would be responsible for even greater changes in labour productivity, then the model shows a decline in the total GDP of the Group 1 economies, even with 'putty capital' (no premature scrapping of existing capital stock). And if the level of scrapping is increased so that only 70 per cent of capital may be switched between production sectors, all economic groups experience a decline in their GDP as a result of the initial adjustment to the new technology.

Figure 7.4: *Distribution and growth versus technological change in the most industrialised countries*

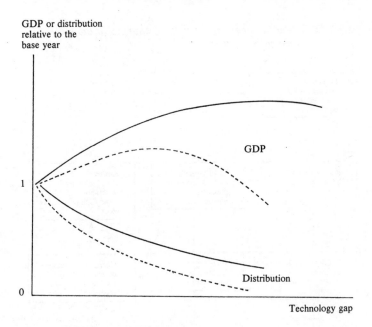

Key: —— no scrapping of existing capital stock
- - - - some existing capital stock scrapped

The implications of the two experiments are shown in Figure 7.4: a country adopting the new technology is, in general, likely to experience a worsening of income distribution as a result of market forces alone. In addition, if the introduction of the technology is very rapid, so that the technology gap between it and other economies widens greatly,

the associated restructuring of the world economy may lead to a
situation where all or most parties in the economy become worse-off,
at least for some time.

What we saw in the first experiment is that the situation of the
economy introducing the technology initially improves at the expense
of other economies. But if the income, and hence the demand for
imports, of those trading partners is sufficiently affected, then the
possibilities for further expansion of production are seriously curtailed.
Any further increases in productivity, then simply lead to a net
displacement of labour in the innovating economy with a consequent
further decline in demand. The processes underlying these changes
are shown in Figure 7.5.

Figure 7.5: *New technology and structural changes in the most
industrialised economies*

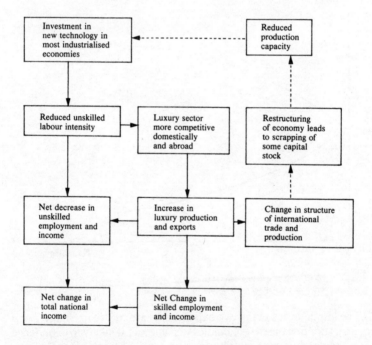

In the second experiment (Table 7.5), a significant fraction of the
total world capital stock has been discarded. If this is not replaced
through new investment, the world economy systematically reduces

in size as a result of the introduction of new production techniques. These experiments suggest, therefore, that a period of very rapid technical change in the most industrialised economies, with liberal international trade and non-interventionist domestic economic policies, could result in an era of restructuring and decline similar to the present world economic crisis. Whether or not this actually happens depends on factors which can at best be treated only in an oversimplified way in the model, but may be contemplated in our scenario building exercise in Chapter 8.

NEW TECHNOLOGY IN THE LEAST INDUSTRIALISED COUNTRIES

In contrast with the two experiments above for the already industrialised economies, we now explore the result of introducing the most modern production techniques into the least developed economies. The question of the availability of the technology becomes especially acute here. Several studies (e.g. Bessant, 1980; Kaplinsky, 1983), suggest that least developed economies may experience particular problems in acquiring and using the technologies. For example, they may lack skills and know-how required to produce or use the technologies.

To examine this change in technology, at least roughly, we assume that the new capital embodying the new technology is employed in the luxury goods sector of these economies. In this case, the overall effect is to reduce employment of skilled as well as unskilled labour, and also to reduce the combined GDP of these economies. With the new technology, the rate of profit rises considerably. Even so, the income of richer households does not rise overall because of the decline in skilled employment. We summarise this result, together with those of other experiments dealing with technical change in the Group 6 economies, in Table 7.6. It should be contrasted with the result of Table 7.2 in which no *explicit* change in technology was made, but which simulated the impact of a modernisation strategy (i.e. a shift of production to the modern luxury goods sector). In that experiment, the employment effects were rather small and the low-income households improved their economic position because of their additional investment income. The comparison here suggests that a *more* rather than a less labour-intensive technology could be advantageous to these economies. Furthermore, the experiment with new technology assumes no additional repatriation of profit. This is rather unlikely under the conditions which would lead to the technology becoming available to these countries. New technology is

Table 7.6: *Summary social accounts for least industrialised economies: The impact of technical change*

Experiment	B0	A2	M2	T2	T3	T5
Output:						
Luxury	95.5	181.4	223.8	313.5	105.7	86.6
Basic	431.5	383.4	349.6	276.6	433.4	476.3
Investment	97.6	80.2	69.9	57.7	121.9	100.5
Factors:						
Skilled	42.5	42.5	39.8	40.3	62.4	49.4
Unskilled	150.6	150.7	141.8	136.9	167.6	199.5
Capital	90.9	94.2	102.1	101.8	61.5	61.1
Households:						
High-income	85.5	85.5	88.1	88.4	65.9	32.6
Low-income	223.3	226.8	220.4	215.4	225.6	277.3
Net trade:						
Luxury	−11	72.1	114.9	204.5	0.1	0.1
Basic	6.1	−55.5	−87.1	−161.3	−0.1	−0.1
Investment	−20	−41.8	−52.6	−67.9	0.0	0.0

Note: Amounts in US$ bn (1975):

KEY:

B0 = Base year 1975
A2 = Installation of new capital/growth of modern sector
M2 = New technology in the least industrialised economies
T2 = Appropriate technology with free markets
T3 = Appropriate technology with closed markets
T5 = Basic needs-oriented technology with closed markets

likely to become readily available only when export-oriented development policies involving the transnational enterprises (as carriers of the new technology) are advanced. Modern-sector technology, by contrast, is already employed in the luxury goods sector and is more readily available to the least industrialised countries under any strategy they choose.

APPROPRIATE TECHNOLOGY AND THE MARKET

What, then, of the introduction of 'appropriate' technologies in the

Group 6 economies? This approach is generally in keeping with a radical or reformist worldview, but has also been proposed in various forms by agencies such as the World Bank. Definitions of what constitutes 'appropriate' technology vary widely, but an extensive survey by Vitelli (1980) showed the most common argument used in support of this technology to be that the more labour-intensive techniques will be competitive with modern-sector production in developing economies. These studies usually assume that factor prices are unchanged by the introduction of the new technology — a standard assumption of neo-classical analysis — and may well be a valid approximation for a specific production unit in a large economy. However, it may not apply across a whole economy in which appropriate technology is introduced extensively as a matter of national policy. After all, a primary objective of introducing these techniques is to *increase* the incomes of at least one of the factors — low-income labour (see e.g. Thirsk, 1981). This change in factor prices obviously could affect the economic viability of the techniques: a technology which is competitive at the prevailing wage level may cease to be so if factor prices change.

In order to use our model to evaluate an appropriate technology strategy, a precise definition of the technology is required. It is possible, from a secondary analysis of existing 'choice of technique'

Table 7.7: *Estimated coefficients for appropriate technology experiment: Factor inputs in basic needs/factor inputs in modernisation strategy.*

	Unskilled output	Skilled output	Labour output	Capital output	Capital labour
Food Processing (sector average)	3.17	0.72	8.33	0.03	0.25
Textiles (sector average)	n.a.	n.a.	4.38	0.48	1.2
Housing (sector average)	1.9	0.05	3.5	0.39	0.32
Averages over sub-sector studies	2.9	0.39	4.0	0.41	0.35
Standard deviation	2.3	0.51	3.5	0.25	0.32

Based on Cole and Nunez-Barigga, 1981.

studies (e.g. Cole and Nuñez-Barigga, 1981), to identify techniques which are compatible with a human needs-oriented development strategy (which we consider in more detail in Chapter 8).

To estimate model parameters for appropriate technology, factor shares were compared directly with those of modern-sector techniques currently used in the production of luxury goods in the Group 6 economies; the comparison is shown in Table 7.7. There are wide variations in the factor inputs within and between sectors, as the standard errors on the results of the study indicate, so that the figures must be treated with caution. However, they are quite comparable to those of other authors (e.g. Bartsch, 1977) and may be seen as a first approximation until more reliable data are available.

Appropriate technology in an open economy

The first experiment with appropriate technology assume that inputs of skilled and unskilled labour and capital to appropriate techniques for the basic goods sector are, respectively, 39, 29 and 41 per cent of those prevailing in the luxury goods sector. The US$30 billion development finance is devoted to these appropriate techniques and added to the existing US$390 billion capital stock already installed in the basic goods sector and owned by low-income households. The aggregate factor inputs are adjusted *pro rata*.

If factor prices do not change, this would ensure that the competitiveness of the sector is marginally increased domestically and abroad, and we would expect production in the basic goods sector to expand. This is not the case, however; the results of the experiment, in Table 7.6, show a rather dramatic fall-off in basic goods production in the Group 6 countries and a much increased dependence of this economy on external sources for these goods.

The reasons for this are as follows. In equilibrium, the introduction of the new technology has triggered a reduction in employment of both skilled and unskilled labour in the least developed economies. The results in a 5 and 10 per cent decline, respectively, in their wages. At the same time, the cost of capital in this economy increases by 8 per cent. A striking feature of these results is that, although the factor price movements are in the opposite direction to those in the modernisation experiment (Table 7.2), the production shifts are in the same direction.

Since a rise of a factor's price is least favourable to those sectors which are intensive in its use it is seen from Table 5.14 that a rise in the price of capital favours expansion of luxury goods production. Similarly, a fall in the cost of low-skilled and skilled labour, respectively, favours basic and investment goods production. Given that production of investment goods is much less capital-intensive than other sectors of this economy, and that the fall in the price of skilled labour is comparatively small, production of capital goods should be favoured; but this does not happen. The explanation cannot be found simply in terms of the prevailing factor inputs: the reason must lie in changes in production techniques or the relationship between production costs in the least developed economy and the rest of the world. The next experiments, exploring the impact of appropriate technology in a closed economy, justify this assertion.

The present experiment indicates a tendency for the combination of appropriate technology and free trade policies to push the least developed economies into greater dependence on other economies, both for basic commodities and investment goods. Furthermore, the increase in the rate of profit in the least developed economy more than offsets the loss in wages of the high-income group. For the low-income group, even the additional ownership of the new capital stock does little to offset their loss of wage income. Hence, the introduction of appropriate technology in these circumstances leads to added external dependence and worsens the absolute and relative incomes of the poorest group.

The figures used for the experiment were extremely tentative, but we can check how sensitive the results are. Halving the amount of skilled labour and capital previously assumed does little to prevent the decline of the basic goods sector. These results support the idea that open economies and appropriate technologies do not mix. (They thus exposes a contradiction in the strategies propounded by US-backed development agencies and the activities of the current US administration.) While their details are very dependent upon a rather insecure empirical base, the experiments do suggest that, within the limits of reasonable estimates of conditions in different countries, this combination of policies is unlikely to achieve the ends desired for it.

Appropriate technology in a closed economy
This incompatability between appropriate technology and an open economy does not constitute an objection to appropriate technology under all circumstances. Is it the nature of the technology, or the openness of the economy, which is responsible for the apparent failure to achieve the objectives set for the appropriate technology? One aspect of a radical human needs-oriented strategy is that an economy should be much less dependent on other economies for critical commodities. This does not imply that it should not trade at all, but

rather that net trade flows for each class of commodity should be relatively small compared with total domestic consumption. (As Table 5.16 showed, this is effectively the case for the more industrialised groups of economies.)

What would be the result of constraining international trade between the least developed economies and the rest of the world? We model this by calculating the equilibrium solution for the Group 6 countries alone, while making the same adjustment to the technical coefficients as for the last appropriate technology experiment. The total net importation of each commodity is not permitted to exceed a nominal US$1 billion. In addition, the balance of payment deficit of the economy is set to zero, so that domestic consumption in the economy is no longer subsidised by transfers from abroad. The behaviour of the rest-of-the-world economy is computed separately.

The result of this experiment, also shown in Table 7.6, provides a major contrast to the results for the open economy. Despite the assumed loss of US$7 billion support (through the balance of payments deficit), the total income of low-income households is above the base year (and also higher than for the 'capital gift' experiment). Unskilled employment rises by some 6 per cent and skilled employment increases by more than 12 per cent above the base year because of the import-substitution of skill-intensive luxury and investment goods. The resulting rise in the wage income of richer households balances the decline in their investment income, but their total income falls because of the loss of international financial support. But, even without this, the income of poorer households is increased above the base year. The introduction of more unskilled labour-intensive technology has indeed increased employment and wages and also facilitated some redistribution.

The experiment supports the argument that the introduction of appropriate technology in an open economy failed to increase employment because of the effects of international market forces (rather than resulting from the structure of the domestic economy or the factor proportions assumed for the new technology). This new experiment indicates that a human needs-oriented strategy, with appropriate technology as a major component, requires restrictions on trade. There are also implications for a developing economy which does employ some measure of appropriate technology and trade restrictions. If it chooses to become more integrated into the world economy, it may experience strong pressures which may unbalance the production structure and worsen the income distribution. This problem may be faced by such centrally-planned countries as the People's Republic of China in Group 6.

DOMESTIC REDISTRIBUTION

A shift in production techniques could also be associated with a change in consumption habits. To explore this possibility, the share of luxury goods in the consumption basket of low-income groups in the least developed economy is reduced by 15 per cent, and that for high-income groups by 20 per cent. Further, since it is likely that a basic needs strategy would also be reinforced by redistributive taxation, it is assumed that US$20 billion of the income of rich households is transferred to low-income households (see, for example, Singh, 1979). The last experiment is then repeated with these changes and the results are shown in Table 7.6.

The domestic transfer of US$20 billion contributes less than half of the additional US$51.7 billion increase in the low-income groups'

Figure 7.6: *Redistribution in a human needs-oriented approach*

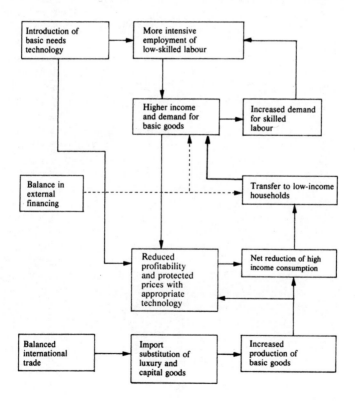

Table 7.8: *Least industrialised developing economies social accounts: Basic needs-oriented technology installed*

	Production sectors			Factors of production			Final demand			Total
	Luxury	Basic	Invest.	Skill.	Unskill.	Capital	High-inc.	Low-inc.	Foreign	
Luxury	4	19	6	0	0	0	1	57	0	87
Basic	39	198	41	0	0	0	11	203	0	492
Investment	10	39	13	0	0	0	21	18	0	100
Skilled	6	20	23	0	0	0	0	0	0	49
Unskilled	18	173	8	0	0	0	0	0	0	200
Capital	8	43	9	0	0	0	0	0	0	61
High-income	0	0	0	49	0	41	0	0	0	91
Low-income	0	0	0	0	200	20	58	0	0	277
Total	87	492	100	49	200	61	91	277	0	1357

Notes: Rows and columns may not add to precise total listed, in this and subsequent SAMs, due to rounding of amounts to nearest billion.

198

budget. The shift in consumption habits also has a significant impact on income distribution through the domestic market. The increased consumption of low-skilled, labour-intensive products by both income classes creates an additional demand for low-skilled labour. This increases the income of poorer households, and in turn raises total demand for the labour-intensive goods, as considered in Chapters 4 and 5. These processes are summarised in Figure 7.6.

These processes come into play because the economy has been protected against external markets. When the composition of consumption of the developing country population changes in an open economy, the structure of trade tends to change. If this happens, the multiplier effect noted above cannot operate since it would imply a rise in the domestic price of labour-intensive goods, which would not be supportable against world prices. The products would be imported, with undesirable consequences for domestic employment.

The social accounts for the Group 6 economies, shown in Table 7.8, throw additional light on the result. The employment and wage income of the skilled labour group rise, together with those of low-skilled labour. The overall reduction of income for the high-skill group is from capital (a loss of profits) rather than from labour (for wages increase). In reality, of course, the high-income group is certainly not homogeneous. This result has quite different significance for capitalists (who live on profits) by comparison with skilled labour (including bureaucrats and management, who live mainly on wages), and the workforce in general. The implied strategy is not, then, attractive to local entrepeneurs, but international development banks and aid agencies might be less adverse to it, since profitability in the recipient economies remains comparable to that in the donor economies.

GROWTH VERSUS REDISTRIBUTION

We have seen that contrasting choices in technology give rise to distinctly different movements in the level and the distribution of the income generated by development finance. In some cases, the total income of the least developed economies falls; in most cases, the income of the high-income group falls, usually because of the decline in the rate of profit as a consequence of the introduction of additional capital. Many of the experiments, therefore, are redistributive; in particular they raise the level of employment and income of the low-skilled group. (The most notable exception is the addition of appropriate technology in an open economy — Table 7.6.)

Given our earlier caveats, it is unwise to make detailed comparisons between the precise magnitudes of the shifts observed. It is, nevertheless, striking that the experiments which could give rise to significant redistribution and potential overall growth in the Group 6 countries — the closed economy strategy and the modern-sector route — have the most conflictual implications for development policy. Both call for significant changes in power relations, as well as in economic strategies. The mechanisms whereby the results come about are also quite different in these two cases. Consequently, the pattern of integration and level of dependence on the rest of the world economy are divergent.

The emphasis on economic growth, as opposed to the redistribution of economic resources, provides one of the principal contrasts between the three worldviews. The conservative view expects that provided overall economic growth is satisfactory, all parties in the world economy benefit. Thus, developing countries should not impede the growth of the richest countries, since, through a process of 'trickle-down', both the poorest people and the poorest countries will benefit. Within countries, the poorest groups will benefit even if no policies of positive redistribution are enacted. The earlier stages of industrialisation may even require that income and wealth are concentrated rather than redistributed, it is argued. The other worldviews place greater emphasis on the creation of mass markets through redistribution or Keynesian-type policies to boost consumer spending. Some authors relate the supposed failure of import-substitution strategies (at least, by contrast to some Asian export-oriented economies) to a failure to redistribute income, and hence to a restricted scope for domestically-generated growth (see e.g. Nayyar, 1979).

The question of growth versus distribution, or whether distribution should follow or precede growth is, therefore, central to the debate between the worldviews and their proposed development strategies. The experiments so far described have been *comparative static* exercises, which compare only the initial effects of alternative policies as an equilibrium is reached. They do not examine the cumulative dynamic effects of these policies. In this section we re-examine some of the earlier experiments to obtain insight into the possible dynamic effects of the alternative development strategies for the least industrialised groups of economies.

One argument for appropriate technologies is that they are less demanding of capital than other technologies, so that a given amount of new investment, in principle, leads to a much higher level of output growth. In the model, levels of output depend on several factors: the total amount of capital stock in each economy; the distribution of production; and the sectoral capital-output ratios. Growth in output can be determined by changes in all these variables. To model these growth processes fully, the temporary equilibrium model used here would have to be made more dynamic. It would be run through a series of equilibria, each representing successive investment periods, and using chosen 'rules' about investment and other behaviour. But if capital stock and factor output ratios only were changed, this procedure would soon run into trouble. Capital would become increasingly cheap, because there would be no increase in the

Table 7.9: *Summary social accounts for least industrialised economies: Growth versus distribution*

Experiment	A2	M2	T5	A4	I2	I5
Output:						
Luxury	181.4	223.8	86.6	210.1	251.4	87.9
Basic	383.4	349.6	476.3	367.3	334.4	482.4
Investment	80.2	69.9	100.5	74.4	64.3	101.2
Factors:						
Skilled	42.5	39.8	49.4	42.5	39.8	51.2
Unskilled	150.7	141.8	199.5	150.8	141.8	205.2
Capital	94.2	102.1	61.1	95.4	103.3	57.1
Households:						
High-income	85.5	88.1	32.6	85.5	88.1	31.3
Low-income	226.8	220.4	277.3	228	221.7	282.2
Net trade:						
Luxury	72.1	114.9	0.1	99.8	141.6	0.1
Basic	−55.5	−87.1	−0.1	−76.1	−106.8	−0.1
Investment	−41.8	−52.6	0.0	−48.5	−59.6	0.0

Note: Amounts in US$ bn (1975).

KEY:

A2	=	Installation of new capital/growth of modern sector
M2	=	New technology in the least industrialised economies
T5	=	Basic needs-oriented technology with closed markets
A4	=	Additional new capital installed/further modernisation
I2	=	Additional investment in new technology
I5	=	Additional investment in basic needs technology

availability of other factors; the production structure would become correspondingly distorted. It is only useful for the purposes of assessing a one-off transfer of development finance to make a comparison neglecting other factors, as we have done above.

Changes in the production structure and the production techniques together should have a significant impact on relative growth rates. To test these phenomena more systematically, allowing for other effects in the model, the earlier experiments can be repeated in turn with a further US$10 billion increase in the US$30 billion additional stock. Production techniques are the same as in the corresponding experiments. The results are summarised in Table 7.9.

First, we repeat the capital gift experiment in Table 7.2 (which simulates a process of modernisation in the Group 6 economies). The additional US$10 billion stock creates an additional US$1.2 billion income for the low-income group. This is approximately *pro rata*, with the impact of the initial US$30 billion which raised low incomes by US$3.5 billion. The increase in total income is thus in line with the earlier experiment.

In the second experiment, the additional US$10 billion investment is associated with the new microprocessor-related technology. Now, the low-income budget falls by US$1.3 billion, and so remains below the base year level.

Last, the additional investment is associated with the addition of appropriate technology (in a human needs-oriented economy). In this case, the distributional impact is reversed, and the gain is more significant — US$4.5 billion. There is, however, a smaller decline in the high income budget. Thus, the low-income group's budget and the total national budget are increased more by the addition of extra investment to the closed economy than they are by further investment in a modernising open economy. In comparing the earlier experiment — between appropriate technology in a closed economy, and new technologies or modernisation in an open, less developed economy — it appears that the former has the greater *potential* for initial growth, and, more especially, for raising the lowest incomes. However, as the structure of the economy changes and levels of employment and wages rise, the most suitable technology for this development will also change. Thus, the choice of technology still has to be viewed as a *dynamic process* (even though it may not always correspond to the Social-Darwinist prescription offered by the conservative worldview).

The outcome of the experiments with the adoption of appropriate technology in a closed economy does not accord with the reformist expectations that development finance should improve the health of industrialised economies. This style of development satisfies even less the criteria for development assistance set by the current (1984) US administration. However, some sympathetic Northern economies (especially in Scandinavia) have been more prepared to support the least developed economies, even when they have followed strictly regulated import and redistributive policies.

It might be argued that there is no reason why the domestic income tax transfer should not be made in the open economy also. But, the implication of the conservative position is that economies which choose to engage in such a massive domestic redistribution are unlikely to be recipients of development finance. The admonition of the *status quo* scenario, that countries should rely on the functioning of the market for economic development, applies to domestic policy as much as to international trade.

By the same token, the idea that the most sophisticated technologies would first be used in the least developed economies to boost production of basic goods for low-income consumption rather than luxury goods for high-income groups has more than a touch of wishful thinking given present patterns of industrialisation. International firms which are the main carriers of the new microprocessor-related technologies are unlikely to locate production in any economy undertaking a politically-motivated redistribution. Hence, our experiments assumed that the new technology is directed exclusively at the more luxury-oriented Northern markets, with any widespread introduction of this technology in the South most likely to involve the import substitution of luxury production. The results suggest that this would lead to a large reduction of the production of basic goods in these economies.

To adapt the technology for production of basic goods in the least developed economies would require access to skills and technology, and a research and development capacity, which few of the poorer economies currently possess. Unless these technologies are provided by like-minded industrial economies, they must remain at best extremely constrained options for the least developed economies adopting interventionist or welfare-oriented policies. From this standpoint, the adoption of modern sector or appropriate technology for basic goods production also has the advantage that they are already assimilated into the least developed countries.

While the experiments cannot provide definitive answers as to the impact of economic assistance or the question of growth versus redistribution, they do challenge much conventional thinking about the most appropriate development strategies to pursue in our volatile world. They indicate some of the strategic trade-offs, new policies and associated coalitions needed if development objectives are to be achieved. Finally, they provide building blocks through which the scenarios set forth earlier can be elaborated — a task to which we now proceed.

8 From Crisis to Crisis — A History of the Future?

INTRODUCTION

In this final chapter we summarise and bring together the building blocks of our analysis — scenario construction and model experiments — to engage in a more complete scenario analysis. We describe one possible future history for the world political economy, up to and beyond the turn of the century. We do not try to predict *the* future here, but merely to elaborate one plausible alternative sequence of events and to derive some lessons for the present from it.

There are two quite different ways we could approach this scenario analysis. One would be to take each of the different strategies in turn, assume that their conditions of realisation are met, and then use the model to inspect the changes these bring about in the world (this would be a simple extension of the model experiments described above). The second, which we shall pursue in this chapter, recognises that the world and the actors in it may move from one type of strategy to another; indeed, the precondition for the realisation of a strategy may be disillusionment with another. Thus, by this second approach, we can add more easily a sense of historical evolution to our study.

Through our scenario we seek to portray a future in which human values, and the institutions which represent them, change. Progress is made in international and domestic relations, both economic and political. The present inexorable trend in income and power begins to change, but this is not achieved without a struggle in which successive crises are confronted, and new global perspectives realised.

CONTRASTING PERSPECTIVES

The contrasting interpretations of world history and the strategies for future change, as we have outlined them, are simplifications of actual

205

proposals. By necessity, our comparative anlysis has tended to caricature the elaborate and subtle prescriptions derived from complex and everchanging worldviews. We classified proposals for the international economy under three headings: status quo, new international economic order (NIEO) and collective self-reliance (CSR). These are contrasted in Table 8.1.

Of the three perspectives, the *status quo* viewpoint (supported by the stronger OECD countries and various other Western international organisations, ranging from the Trilateral Commission to international banks) is the most sanguine about current economic trends, and is most prone to interpret the present world economic crisis as largely an unfortunate conjuncture of unusual occurrences. Generally, its proponents advocate an extended international division of labour, with the most advanced economies providing the dynamic of worldwide economic growth. New technologies (such as microelectronics and biotechnology) and perhaps the economic exploitation of outer-space and ocean resources, are seen as central to this revitalisation of the world economy, and transnational firms and global corporations as the major agents for the diffusion of new products and processes. Some temporary interventionist policies at the domestic and international level by governments may be needed to guide this technological revolution, but, to a large extent, world and domestic markets offer the most appropriate guide to an efficient and optimal international division of labour.

By contrast, advocates of NIEO proposals see the present crisis as of greater magnitude, as requiring a series of international mechanisms to ensure that world markets operate fairly and to reduce instabilities inherent in the present system. The view (favoured by leading countries in the Group of 77 and important political forces in several Western governments, especially a number of the smaller countries) is also more interventionist in respect of domestic policy, more welfare-oriented at the national and international level, and generally more internationalist. New production techniques are seen as posing a possible threat (through loss of employment), as well as opportunities for expansion (through the creation of new demand). This view favours systematic financial assistance to developing countries, with proposals such as those of the Brandt Commission, for a new Marshall Aid-type plan for the Third World. At the very least, these proposals require that world markets be no longer structured in favour of the industrialised nations and their transnational enterprises under the guise of free trade.

Most NIEO proposals start from a view of the world political

Table 8.1: *Dominant policy assumptions underlying alternative strategies for the international economy*

	Status quo	New international economic order	Collective self-reliance	Human needs-oriented
Principal assumptions	All regions gain from international trade. Growth in the North leads to growth in the South. Productive and innovative forces best stimulated through the market with a minimum of interference from national or international administration.	Status quo is economically and politically unstable. Some parts of the South and East have increasing power and some international restructuring and concessions must take place. Underlying economic principles much as status quo although more government and intervention is needed to avoid instability and achieve welfare objectives. Redirection of arms spending would reduce nuclear threat and increase productive growth. Growth in South can assist growth in North.	Difficult or impossible to achieve non-dependent growth in the South within present structure of international economic and political relationships. Many internal structural problems in the South are because of links with the North. South is quite capable of developing within more self-reliant strategy. Crisis in the North will deepen and exacerbate Southern problems.	Development should concentrate on satisfaction of basic material and social needs. Many skills in less developed countries (LDCs) are under-utilised and should form a starting-point for a basic needs-oriented strategy. Trade, aid, and transfer of production and technology are not always beneficial to least industrialised economies.

207

Table 8.1: *Continued*

	Status quo	*New international economic order*	*Collective self-reliance*	*Human needs-oriented*
Global strategy	Increased specialisation in the international division of labour. Advanced economies and sectors could provide dynamic for growth in the world. International stability can be advanced through balanced arms expenditure between major powers.	International welfare and disarmament policies. Systematic assistance to developing countries especially transfers of production and technology to South. Some element of global planning, e.g. aim to meet Lima Declaration targets for industry and finance.	Systematic delinking of the South from the North with collective action to promote Southern development and reject continued dependence from the North or attempts at recolonisation. Northern countries increasingly adopt mercantillist policies as a result of failure of liberal policies.	International links should be re-examined and restructured to further national domestic basic needs requirements. Production systems should be structured so that through the market a more equitable distribution is achieved. Aim would be to create effective demand among low-income groups so as to reorient international links to their needs.
Future perspective	Overall optimal world economic system with maximum growth. Development may be uneven but without disastrous military conflict or environmental damage.	More equitable and stable growth worldwide with decreasing international tensions.	North will become increasingly multi-polar, protectionist, militaristic and crisis-ridden. South will build sufficiently strong and independent institutions and more production systems more relevant to needs. Growth will shift to the East and South.	Overall growth levels should not be seriously affected. Internal distribution of income and control improved. Provides a basis for long-term sustained growth.

economy as in extreme crisis — or even as rapidly approaching major disaster, which can be averted only through internationalist global reforms.

Advocates of the CSR position agree with these points, but question whether the industrial countries would, in practice, comply with any international agreements of an NIEO. They see the crisis in the industrial economies as being deeply rooted, and sense in this both an opportunity to mobilise forces for change, and a need for countries to adopt a more self-reliant approach. Proponents tend to be extremely critical of many existing institutions, and are thus more often members of independent groupings than of governments. Nevertheless, this strategy has advocates among Third World governments, political parties, and in international organisations. They argue that many structural problems in the developing nations are the result of earlier colonial practices and the present manipulations of the industrial powers, and consequently, they look for a more independent development path by Southern countries as a whole.

THE RELEVANCE OF THE MODEL EXPERIMENTS

Chapter 7 describes three sets of experiments relevant to these contrasting perspectives, dealing with the possible impact of development aid and foreign investment, the choice of technology, and the trade-offs between economic growth and distribution under different assumptions about technology and international exchanges. These experiments give insight into possible consequences of different global policy measures.

The results from the experiments on development aid suggest that when a developing region receives aid, the benefits may be redistributed worldwide in such a way that the recipient economy gains relatively little additional income and little reduction in its poverty level. Market forces create an economic environment in which production structures and international trade-flows are liable to adapt to relatively small changes in the level of income. Thus considerable social disruption, such as that associated with migration to urban areas, could be engendered without compensating increases in material well-being or redistribution. This result applies whether the development finance is transferred as a gift, a loan, or as direct foreign investment. Consequently, the results are relevant for those perspectives which propose development aid to further social policies.

The functioning of markets is such that development finance may

in some cases have a greater impact on the income distribution of the
donor (and other) economies than on the recipient. Even so, contrary
to the assumptions underlying the NIEO perspective, development
finance may be only marginally effective in bringing idle capacity in
the Northern industrial economies back into production.

The second set of experiments deals with questions of technical
choice. Three possible directions for future technical change are
explored. The first is the assumed implementation of labour-displacing
new microprocessor-related technologies in the leading sector of most
industrialised economies. In effect these economies would be responsible for leading the world economy into a technological revolution, as
promoted by conservative and some reformist strategies. The results
of the experiment show that while this process would improve the
relative position of the leading economies in the world, it would also
worsen their domestic income distribution. In other economies there
are both positive and negative effects on both income and income
distribution.

The model's results show the possibility of a progressive spiral of
decline being triggered off by dramatic changes in technology. Large
or rapid changes in technology lead to the obsolescence of existing
plant both domestically and abroad, as a result of which the overall
gains to all economies are likely to be reduced. Whether a decline in
productive capital with a slowdown in overall economic growth
(exacerbating the effects of the present world economic crisis) is
consistent with present trends cannot be determined unambiguously
with our data. But the results suggest that changes in technology
worldwide should be coordinated with other policy changes in the
world economy if desirable global objectives are to be reached
without great human cost.

The second experiment looks at the introduction of new technology
in the least industrialised economies. This might follow, for example,
the opening-up of the economies of these countries and a rapid
diffusion of technology through the operation of transnational enterprises there. The results again show a worsening of income distribution.
Further, although the rate of profit in these economies increases
markedly with the introduction of the new technology, overall income
appears to decline.

The third experiment considers the introduction of so-called
appropriate technology into the least industrialised economies. Two
cases are considered: (i) with these economies exposed to international market forces; and (ii) with a more autarkic style of
development. The conditions placed on international trade are shown

to be a major determinant of the impacts of technological choice. When a less industrialised country opens its economy and adopts (unskilled) labour-intensive appropriate technology in order to maximise employment, the results, at least from our model, appear disastrous. But if the economy is more or less isolated from international markets, appropriate technology can serve a very significant internal redistributive function. This suggests that *if* the destabilising effects of external market forces on local production systems can be brought under control (through a tightly regulated trade policy, for instance), then domestic market forces can actually reinforce social policies. Increased domestic demand leads to increased employment and wages, and so to further increases in demand. In some cases, the loss of some national growth, which would otherwise have been gained from international trade, can be compensated for by feedback effects arising from these cyclical flows of income in the domestic economy, which reinforce other social objectives.

The results of these last experiments, together with those on development aid, suggest that for the least developed economies pursuing a policy of free trade, the modern-sector technologies are likely to promote best the objective of economic growth. The objective of improved incomes for the poorest households could also be satisfied, although this would be at the cost of increased dependence on the outside world for staple commodities and technology. Although the use of appropriate technology could mean the sacrifice of some overall growth, the apparent growth advantages of an open economy using more sophisticated techniques might well be equally provided by a closed economy using appropriate technology, due to changes in sectoral productivity and the composition of output.

These results depend on a small number of central assumptions, which may be felt to be more or less plausible representations of the global political economy. These are:

(1) the existence of world markets for commodities and domestic markets for factors of production;
(2) differences in factor inputs between sectors: labour intensity in all sectors is generally highest in the least industrialised economies, and capital goods production is especially skill-intensive in all economies;
(3) differences in the availability of factors between economies: abundant unskilled labour in the least developed economies and abundant skilled labour and capital in the most advanced economies.

Our model has a simple structure, and it is to be expected that its results will sometimes be exaggerated. They are relatively insensitive to some parameter values and assumptions, and rather sensitive to others. The precise magnitudes of some parameters are, therefore, important, and we do not always have adequate data. However, even if the estimation and structure do not provide a perfect reflection of each economic group, they may well still capture the situation of many economies within the groups.

The results of the model have socio-political, as well as purely economic, implications. At the domestic level, the redistribution of income which occurs between capital and labour, or between high- and low-income households, for instance, will affect the political conditions under which subsequent policies could be enacted. The results from the model also show that the uneven benefits of particular strategies often cut across national boundaries. Indeed, in some cases the national economic interest seems weaker than the economic interest shared by sub-national actors across different economic groups. However, we should bear in mind here that the possibility of acting on such potential international coalitions may be blocked by existing political structures. In this respect, the relationships and aggregations in the model are necessarily highly simplified. There are considerable differences even within the economic groups defined in the model — for example, in the attitude of various industrial countries towards Third World development. Thus, in Group 1, differences between the USA and some European governments in the support awarded to different types of Southern development strategy may, in practice, prove to be very significant.

Thus, while our results may suggest important tendencies in the world system, they derive from a set of restricted formal categories and rigid relationships. In combining the results of the model with the strategies suggested by the worldviews, these limitations have to be faced and, where possible, overcome. In order to achieve greater flexibility, in the following scenario analysis we relax the rigid behavioural relations in the model (as employed in Chapter 7) and now use the Social Accounting Matrix framework principally as a way of preserving overall consistency when we represent scenarios in a quantified manner.

The scenario analysis will provide qualitative and quantitative descriptions of a particular future history. The magnitude of some key variables, such as growth rates and the composition of international trade, are imputed from the verbal description of the scenario, the expectations of the international agencies reviewed in Chapter 1, and our model experiments. These are used as starting-points for the preparation of new sets of social accounts. We thus attempt to go beyond the very impressionistic description of growth rates as 'high' or 'low' employed in our previous work (Freeman and Jahoda, 1978), without falling into the modellers' trap of implying that the quantitative results give a precise picture of any global development path.

A POSSIBLE FUTURE HISTORY?

The formulae for global development provided by the worldviews are ideals, and not all actors in the world economy share them at any one time. However, conditions may arise which bring about greater coherence in policy — nations may emphasise one strategy rather than another; governments may fall, or realise the failure of their existing policies. In effect, new approaches to development and, more especially, the implementation of new approaches, arise out of the failure of present strategies. When these failures erupt into major crises, there may be significant branching points in world history, following which a new regime of dominant actors and policies can arise, transforming or replacing the prevailing worldview. In this century, such branching points may be seen in the two world wars, and the liberation struggles in some Third World countries.

For new policies to be considered and enacted, a body of actors capable of effecting change must share a common perceived interest in that direction of change. However, excluding major wars and anti-colonial struggles, this unity of purpose has rarely occurred on a world scale. The formation of the League of Nations after the first world war, and the United Nations after the second, were major steps towards a more global ideal.

While some world regions have witnessed long periods of relative calm, in the wider view, history has seen a world passing from crisis to crisis. We see little reason to expect that this will not also apply to the history of the future. The period of sustained growth up to the early 1970s was, after all, quite untypical by historical standards. The vast number of international and civil wars in the recent past suggest that the future is likely to see a rapid succession of major changes in direction. Crises of different magnitudes may bring these about for at least a part of the world's population.

Possible futures may be likened to a tree where each branch points to alternative choices in the overall path of world development. This tree-like representation of events is familiar to proponents of

risk analysis or decision theory, where each branch is attributed a
certain probability of occurrence. The sequence of events that we
describe below is only one set of possibilities drawn from many
alternatives. We do not try to place a figure on the likelihood of this
particular sequence of events arising. These possibilities represent
one plausible future, mapped out from our scenario construction; and
it presents some stark choices which we would do well to be prepared
for. Crises which shatter the old order are unpredictable in their timing

Figure 8.1: *A possible future history 1980–2020*

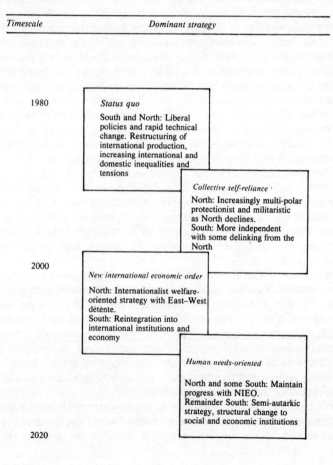

Timescale	Dominant strategy

1980

Status quo

South and North: Liberal
policies and rapid technical
change. Restructuring of
international production,
increasing international and
domestic inequalities and
tensions

Collective self-reliance

North: Increasingly multi-polar
protectionist and militaristic
as North declines.
South: More independent
with some delinking from the
North

2000

New international economic order

North: Internationalist welfare-
oriented strategy with East–West
détente.
South: Reintegration into
international institutions and
economy

Human needs-oriented

North and some South: Maintain
progress with NIEO.
Remainder South: Semi-autarkic
strategy, structural change to
social and economic institutions

2020

or detailed consequences, and we do not place particular weight on the timescale chosen; for convenience in portraying the events and presenting the corresponding results of the model, we follow each new dominant policy for ten-year periods. The timescale for each could be five or even twenty years. As well as being warned that what follows is not prophecy, readers should be aware that this neat segmentation of future history disguises the fact that policy regimes inevitably overlap. Transitions between periods are fluid things – recall our discussion of periodisation of the post war period, in Chapter One.

The sequence of events we follow is illustrated by Figure 8.1. For the next few years the (*status quo*) approach of the late 1970s and early 1980s is maintained, with the world economy stumbling further along its present course. This period may witness a degree of relief from the present recession, a partial recovery from which some countries may substantially gain during a process of economic restructuring; but overall protectionism and militarism increase as national tensions heighten and international inequalities are enlarged. Beyond this horizon few people would care to make predictions with confidence.

Quantitative assessment of this first epoch, up to 1990, essentially involves judging the likely outcome of present trends. This can be based on studies such as those mentioned in Chapter 1, but we are also taking account of the model experiments and our interpretation of the theories described in Chapters 2–4.

Following this era, until the turn of the century, we speculate that a new phase of development is consolidated in the South. For a combination of reasons — 'economic' factors like lack of demand, or closed markets in the North for developing countries' products (including oil), and 'political' factors like outrage about Northern military involvment in Third World affairs — a significant number of countries of the South come to rely increasingly and more explicitly on each other. An inward-looking phase of collective self-reliance is inaugurated in Third World regions.

Although this strategy is threatened to some extent both from within and without, it builds political and economic institutions which enable it to become an important underpinning for the subsequent era of global reform. Economically, the strategy worsens the situation of the North, whose export markets are further reduced. The benefits to the richer developing economies too are limited: These countries are unable to develop domestic mass consumption rapidly by means of high growth or major income redistribution. They do, however, tend to replace the industrial economies in terms of their economic and political dominance over the least industrialised groups.

For the Northern countries, a critical point is finally reached in the late 1990s. This crisis takes place at many levels — escalation of tensions and military build-up between the Nato and Warsaw Pact powers continues (and exacerbates actual confrontations in all regions of the world). The rising protectionism between the market economies (and in a less overt way, within the Soviet bloc) reaches proportions reminiscent of the 1930s, and tensions build up within economic groups as well as between them. Within many countries in the North, levels of unemployment and economic hardship increase to intolerable levels. The choice of global war or global cooperation is finally confronted: influential countries draw back from the brink, and the crisis is partially resolved according to reformist NIEO principles. This involves, at the international level, new institutions, substantial development aid to the Third Word, and serious disarmament efforts worldwide.

In this new epoch, economic advantage is conveyed by innovative industrialisation. For many countries there follows a phase of high growth, reminiscent of the post-war boom. But the newly-establishing world order is threatened by national self-interest, which the more enlightened international reforms have masked rather than abolished. In the North, there is a realisation that the major benefits are accruing to the industrialising, developing countries. While the latter countries are able to exploit low wage costs and to repress emerging political opposition, some of the declining industrial powers, with rigid social structures, find previously secure portions of their economies challenged from new sources of external competition.

Many of the least industrialised economies would still remain marginal to the world economy and so gain little from the new arrangements and economic policy orientations – apart from arrangements for the alleviation of famines and other emergencies associated with mass poverty, these mainly cater for the more rapidly growing and powerful newly-industrialised countries. Politically, there is thus some retreat from the more internationalist perspective and a tendency for military and ideological alliances to reform. Nevertheless, the new institutions which have evolved can withstand and accommodate to the worldwide restructuring of global power relations which is now underway.

At this point, many of the least industrial nations are faced with internal disruption, and are forced to recognise that the prevailing economic and political international framework still places impossible constraints on their domestic development. Unable to influence this international framework, they partially withdraw from it, less in a spirit of confrontation than one of desperation. They begin to turn to autarky and regionalised self-reliance, with human needs-oriented development strategies. However, this is not at the cost of all goodwill from the richer nations: unlike previous attempts to pursue alternative development strategies, in this case, systematic large-scale assistance is afforded to these nations from the international community. This has an initial positive impact on malnutrition and poverty. Although only a first step, it proves decisive in establishing a longer-run equitable development path for the poorest economies. Eventually, these nation begin to feel that the costs of isolation outweigh its benefits, and be gradually reintegrated through appropriate institutions into the world economy.

We shall follow through this future history in greater detail. Even so, we must necessarily summarise most of our account. For each phase of development we give a tabular summary of the dominant policy orientations with respect to key policy issues, such as income distribution, international relations, and choice of technology, and the major forces which tend to undermine the policy. These tables also set out a number of issues including the composition and location of domestic production, the emphasis given to education and vocational training, and to social policies, and possible trends in private life-styles. For each economy group we summarise the policies implemented and the processes taking place in their domestic economies; and describe the transformations taking place in international political and economic institutions (including transnational firms). Because of the inevitable constraints of space, these summaries cannot be elaborated in detail here: they are simply intended to point to the wider implications of the scenario.

Table 8.2 (a): *Annual growth rate guidelines for major scenario variables: period 1975–80: recognition of the crisis*

Economy group	1	2	3	4	5	6
GDP	3.5	3.0	4.5	5.0	4.5	4.5
Population	0.6	1.0	0.9	2.6	3.4	2.1
Productivity	5.5	5.0	6.0	5.5	4.5	4.5
Domestic transfers	–1.0	–1.5	0.0	–0.5	–0.5	0.0
International transfers*	–89.5	9.7	19.4	29.1	2.9	29.1

Table 8.2 (b): *Period 1980–90: Worsening worldwide economic crisis*

Economy group	1	2	3	4	5	6
GDP	1.0	0.5	1.2	2.0	2.0	1.5
Population	0.4	0.8	0.7	1.8	3.0	1.8
Productivity	4.0	3.5	4.0	4.5	4.0	2.0
Domestic transfers	−1.5	−2.0	2.0	−1.0	−0.5	−1.0
International transfers	−72.7	1.1	10.7	31.1	6.4	22.5

Table 8.2 (c): *Period 1990–2000: South strengthens as crisis in North continues*

Economy group	1	2	3	4	5	6
GDP	0.5	0.5	0.5	3.0	2.5	2.0
Population	0.4	0.8	0.7	1.5	2.5	1.0
Productivity	2.0	1.5	2.0	3.0	2.5	2.0
Domestic transfers	−1.5	−2.0	−2.0	−1.0	−0.5	−1.0
International transfers*	−50.0	5.0	10.0	21.0	−5.0	19.0

Table 8.2 (d): *Period 2000–10: First decade of the new international economic order*

Economy group	1	2	3	4	5	6
GDP	2.5	1.5	2.0	4.0	4.0	3.5
Population	0.3	0.5	0.4	1.2	1.8	1.2
Productivity	5.5	3.5	4.5	5.0	5.0	4.5
Domestic transfers	2.0	0.0	2.0	1.0	1.0	1.0
International transfers*	−130.0	−10.0	20.0	60.0	10.0	50.0

Table 8.2 (e): *Period 2010–20: NIEO continues with human needs-oriented strategy*

Economy group	1	2	3	4	5	6
GDP	2.5	2.0	2.5	4.5	4.5	6.0
Population	0.2	0.3	0.2	0.5	0.8	0.5
Productivity	- 5.5	5.0	5.0	5.5	5.0	1.0
Domestic transfers	3.0	3.0	3.0	2.0	1.5	2.0
International transfers*	–156.0	–12.6	–11.6	29.4	38.9	112.8

* *Note*: all data are percentage growth rates, except international transfers, which are US$ bn at end of period.

Table 8.2 gives the assumed average growth rates for some of the major scenario variables during each of the five time-periods. These are based on our qualitative judgements for the different phases of the future history we are describing. In each time-period, we calculate a new set of social accounts for the six economic groups, and the trade and financial flows between them. Further details of the social accounting matrices for each economy group are also shown (these may be compared with Table 5.16 for the base year). For comparability we keep the proportion of high- to low-income households in each economy the same as the base period. Consequently, as the labourforce becomes progressively more skilled, skilled as well as unskilled labour begins to contribute to the income of low-income households.

We now review this stream of events as they may appear in hindsight.

THE CRISIS CONTINUED: 1980–90

The era of worldwide economic crisis which continued up to the mid-1990s involved the increasingly frustrated attempt to implement the *status quo* principles summarised in Table 8.1. The period had two principal phases. The first had begun before 1975 and drew to a close around the mid-1980s. (The situation of the world economy in 1980 is given in Table 8.3.) During this phase, a process of restructuring took

place in which some economies — notably the stronger Groups 1 and
4 — benefited, while the weaker industrialised Group 2 economies
declined. There was also continued growth of private sector transnational institutions. The oil-exporting economies, too, experienced
high rates of economic growth, although increasingly lower than in the
boom years of the 1970s.

Table 8.3a: *Summary social accounts for all economies: worsening
worldwide economic crisis at 1980*

Economy	1	2	3	4	5	6	World
Output:							
Luxury	3289	1275	1230	340	165	96	6395
Basic	1176	554	891	538	215	527	3902
Investment	1161	522	546	223	94	114	2659
Factors:							
Skilled	818	457	211	134	86	62	1769
Unskilled	1263	527	462	334	124	175	2884
Capital	624	312	681	224	94	100	2035
Households:							
High income	971	505	546	249	71	92	2433
Low income	1644	801	828	472	236	275	4256
Net trade:							
Luxury	27	−14	−15	−24	40	−14	0
Basic	−12	−11	10	15	−12	10	0
Investment	75	15	−15	−20	−31	−25	0

Note: Amounts in billions of 1975 US dollars

The less dynamic and less industrialised economies were, however,
only the first casualties of the deepening worldwide economic crisis.
The collapse of demand in these contributed to a progressive spiral of
decline throughout the world economy. Some significant fluctuations
in growth rates arose as nations gained a temporary advantage
through innovation or policy; and there was even some lessening of
the coordination of economic cycles between economies as protectionist policies grew. But this provided only partial relief, and, as the
situation worsened, both public and private policies became increasingly short-term and ameliorative.

Table 8.3b: *Summary social accounts for all economies: worsening worldwide economic crisis at 1980*

Economy	1	2	3	4	5	6	World
Output:							
Luxury	58	54	46	31	35	13	49
Basic	21	24	33	49	45	72	30
Investment	21	22	20	20	20	15	21
Factors:							
Skilled	30	35	16	19	28	18	26
Unskilled	47	41	34	48	41	52	43
Capital	23	24	50	32	31	30	30
Households:							
High income	37	39	40	35	23	25	36
Low income	63	61	60	65	77	75	64
Net trade:							
Luxury	1	–1	–1	–7	24	–15	0
Basic	–1	–2	1	3	–5	2	0
Investment	6	3	–3	–9	–33	–22	0

Notes: Data are percentages
Net exports are percentage of output by sector

The major policy orientations during this period are summarised in Table 8.4. This also depicts the tendencies which eventually undermined the strategy, and were responsible for countries breaking away from it on a substantial scale.

The characteristic policies and behaviour of the various economic groups and international institutions in the early phase are outlined in Table 8.5. These patterns changed so dramatically that by the end of the *status quo* regime they came increasingly to resemble those characteristic of the next phase of development. The situation among the Northern economies became increasingly tense, but this did not yet provoke dominant political actors to re-evaluate the terms on which they approached each other and Third World problems. In particular, there was a failure to make significant compromise on the increasing debt burden of the South — other than temporary adjustments to ward off major bank collapses — and the flow of public and private finance to the developing countries became permanently negative. Nor was there any substantial attempt to negotiate long-term agreements with countries depending on income from cash crops and raw materials.

Table 8.4: *Dominant policy orientation and adverse tendencies in the status quo regime*

	Policy orientation	*Adverse tendencies*
Distribution, welfare and aid	Markets to determine 'trickle-down' or growth between and within nations. Aid mainly for disaster relief and to stimulate new markets.	Basic needs in many South and some North countries are not met for proportion of population. Aid is used to enter markets and as face-saving activity for repressive government and international system.
Life-style	Consumption patterns set by most advanced economies through demonstration effects and stimulation of effective demand.	Unemployment and dualistic/underground economy develops. Disruptive effects of perceived inequalities.
Urban/rural	Industrial and industrialised services growth as dynamic for economies. Modernity implies urbanisation. Agriculture hould respond to world markets.	Increasing neglect of agriculture and use of agriculture surplus to subsidise urban industry, together with improved productivity in North limits growth of basic agriculture and rural industry in South.
Technology	Technology should be market-oriented. A labour- and capital-saving technological revolution is to the general advantage, and any difficulties will be temporary. Traditional techniques and products phased out as informal sector is integrated into the modernising economy. Major sources of new technology (including appropriate technology) are large firms in the North. Entrepreneurship stimulated by removing barriers to trade and innovation. Environmental restrictions minimised, controls on monopolistic practices and trade unions.	Internal and international distribution and consumption patterns would be manipulated to maintain technology and import-dependence on advanced economies. Technology would not reflect factor prices or optimal scale of activity in less advanced economies. Major patents in technologies relevant to South are pre-empted by MNCs. MNCs control transfer of technology between developing countries.

Table 8.4 *Continued*

	Policy orientation	*Adverse tendencies*
Finance	Investment on economic terms at individual project level so as to achieve uniform rate of profit between sectors and regions. International banks controlled by North with some Group 5 participation. Private finance plays an important role.	Private finance is piecemeal and opportunistic and does not reflect national needs but is used to influence internal politics. International banks would apply strict monetarist criteria and support export-oriented industries and mainly high-income groups in South. International debts would increase and difficulties in repayment increase.
Skills and education	Innovation in North. Productive skills developed in South through Northern-oriented education systems and modern industries led by MNCs.	Local talent in LDCs drawn into MNCs and traditional skills neglected.
Raw materials	Short-run views with conservation only when an economic necessity; some stockpiling for strategic and stability reasons. Freedom to exploit oceans, outer-space.	Both industry and stockpiles controlled by dominant firms. High rate of extraction calls for increasingly sophisticated techniques from North and may not be optimal use of resources by South.
Transnational enterprises	Major source of innovations and expertise. Essential to transfer of know-how and production to South.	Effectively controlled in interests of capital. No allegiance to national needs.

Table 8.4 *Continued*

	Policy orientation	*Adverse tendencies*
Orientation of production	No special systematic attention to basic goods production or technologies. Optimal strategy is to rely on market to generate growth at all levels.	Basic industry and agriculture would be systematically neglected in less advanced regions as effective demand and hence investment and technical change are concentrated in luxury and capital sectors. Adoption of increasingly sophisticated technologies and products orients production structure and individual products towards luxury goods consumption.
Employment	Full employment not an objective *per se*. Need to maintain occupational mobility. Factors used according to optimal profitability.	Increasing and even massive unemployment in LDCs and many Northern economies. Many valuable skills wasted or under-employed.
Trade	Trade encouraged and export-led policies including setting-up of export enclaves.	Export-led growth often against interests of weaker nations both in North and South because of impossibility of achieving equitable terms of trade or because of market effects.
Institutions	Functional linkages building on existing institutions.	World institutions and personnel reflect Northern interests.
Population	Explicit population control measures advocated in developing countries.	Continued cycle of poverty and population growth in poorest countries. Lack of purchasing power remains principal barrier to human needs satisfaction.

Table 8.5: *Status quo scenario for economic actors*

Group 1 — Production system and products become increasingly sophisticated. Impact is transmitted worldwide through trade. technology transfer and demonstration effects in consumption. Differentiated but high rate of introduction of labour-displacing techniques but increased competitiveness gives increased trade, employment and domestic incomes a high rate of profit and investment, etc. Systematic changes keep other regions technologically dependent. Change in demand structures reduces relative level of demand for imported basic consumption goods. Arms spending and international tension kept at a level which suppresses productive investment in Group 3. Increasing BOP surplus.
Outcome: Initially this region does well in an increasingly competitive scenario although ultimately, as world demand falls off, there is a pathological conclusion and pattern in Group 1 tends to that in Group 2.

Group 2 — Group declines and loses competitiveness in technologically-advanced capital and luxury goods to Group 1 and price-competitive products to Group 3 and Group 4. (Low demand, low profitability, low wage levels and increases in productivity are insufficient to make products competitive, low rate of investment leads to low average wage of technical change.) Changing technology leads to reduced employment not offset by increased demand. Demonstration effects increase demand for imported luxury goods from Group 1. Worsening income distribution, reduced welfare payments and public expenditure on basic goods. Production of luxury goods increasingly requires high proportion of capital goods imports.
Outcome: Systematic cycle of decline in these economies throughout the period with increasing periphery characteristics (e.g. technological dependence, export enclaves). Economies increasingly overtaken by Groups 3, 4 and 5 and adopt increasingly protectionist and inegalitarian policies with increasing domestic unrest. Structure moves towards Group 4.

Group 3 — Increasing integration into world economy but difficulties in obtaining latest products and technologies from Group 1. Basic-oriented production strongly subsidised by high prices for increasing (demonstration effect) demands for luxury imports and import substitutions products. Basic export levels and prices sufficient to finance imports. Some pressure on income distribution and consumption to finance capital goods imports and armaments. Increasing financial and technological dependence on Group 1 leads to vulnerability by end of period. Increasing competition with Group 1 for imports of raw materials from South.

Table 8.5: *Continued*

	Outcome: increasing tendency for structure to reflect Group 1. Overtakes Group 2 in some sectors but continues to lag behind Group 1 and follows general slowdown trend in world economy.
Group 4	Exports of manufactures to Group 1 and especially Group 2 drive this economy. Rapid technological change in Group 1 reduces demand until Group 4 introduces comparable sophisticated techniques. High-income groups consumption patterns link to Group 1. Low incomes rise slowly or are held down with fewer welfare payments. Rapid technical change in import substitution and export luxury and capital sectors. Less change in basic sector because of lower effective demand and investment. Continued growth of TNC interests and technological dependence, increasing international debt. *Outcome*: Continued inegalitarian growth but systematic improvement in international ranking.
Group 5	Economy driven by world demand for fossil fuels. Especially high rate of profit in this sector used to modernise all sectors including basic. Increasing luxury consumption for high and low incomes. Considerable and wasteful expenditure on non-productive luxury and capital projects. Rate of resource exploitation initially above domestic financing needs. Foreign investment especially to Group 1 (zero net flow) and some to Group 3. Subsidies to manufactured exports and domestic industry needed to make competitive with low income Group 4 and efficient Group 1 but ultimately these are not competitive. *Outcome*: Overall a high but unequal growth in incomes even with high arms expenditure and wasteful consumption, although danger of not industrialising successfully (i.e. joining Group 4 before resource base exhausted).
Group 6	Economic growth determined by world demand for falling and increasingly non-competitive basic sector. Small high-income sector based on enclave and plantation economy. Links to Group 4 and North consumption patterns creates demand for luxury imports and some capital goods for import substitution and extraction industries. Subsistence sector failing and increasingly exploited. Worsening employment prospects for low-income unskilled workers even at subsistence wage. Exports of basic goods result in worsening domestic effective demand. Any aid or technical assistance absorbed by formal high income sector. *Outcome*: Continuation of dualistic economies with subsistence sector not integrated and declining.
International institutions	Positive role of intergovernmental agencies declines and emphasises conflict prevention and ameliorative function. Increasing conflict within agencies presages partial Southern delinking. Influence of private sector on governmental agencies increases and TNCs are carriers of technology and finance (via off-shore processing and international banking system).

The strategy led to increasing divergence in the relative growth rates of different types of economy. There was a further strengthening and concentration of the transnational sector worldwide, and global corporations were accused of owing no allegiance to host economies. The 'enclave' characteristics of developing economies were extended (with more free trade zones and the like). Uneven technological development worsened income distribution in the less dynamic industrial economies of Group 2: unemployment increased and there was a relative worsening of unskilled wage levels.

There was also some weakening of the links within the bloc of centrally-planned industrial countries. The USSR, in particular, was increasingly unable simultaneously to support high defence spending and new industrial capacity, including that which might provide relatively inexpensive raw materials for other members of Group 3. There was a divergence of interest comparable to that within the market economies of Groups 1 and 2, and discrepancies in economic and technological standards widened further. External political circumstances still held the bloc together, but internal political developments made its cohesiveness increasingly fragile. However, despite these troubles, the industrial Soviet bloc countries fared relatively better during this time than their more market-oriented counterparts.

The early successes of some of the newly-industrialised countries in Group 4, and the oil-exporting economies in Group 5, were systematically eroded, as protectionist policies and lack of demand reduced the possibility of export-led growth policies. This was exacerbated as some large Group 6 countries also attempted to export quantities of industrial products. But these latter countries experienced an additional difficulty (indicated in the model experiments in Chapter 7). As previously closed large economies (such as China), which had relied on less advanced or appropriate technologies for production of mass consumption goods, became more fully exposed to world market forces, they faced pressures on their production structure and income distribution. Rapid modernisation had been intended to provide a basis for more independent development within the world economy; but this was less successful than planned, due to lack of funds, of access to the necessary technology, and of skills and training appropriate to the technology that was acquired.

The outcome of these difficult years is reflected in Table 8.6. By comparison with the base year (Table 5.16), by 1990 the share of wages to skilled workers in all economies increased. The share of household income to richer families also increased in all but the least

Table 8.6(a): *Summary social accounts for all economies: South strengthens as crisis continues in North beyond 1990*

Economy	1	2	3	4	5	6	World
Output:							
Luxury	3663	1390	1435	463	201	117	7268
Basic	1235	558	959	619	259	608	4239
Investment	1327	523	608	266	118	132	2974
Factors:							
Skilled	1136	601	320	218	129	85	2489
Unskilled	1228	493	477	365	138	206	2908
Capital	623	268	730	260	103	100	2084
Households:							
High-income	1357	710	717	359	121	106	3370
Low-income	1557	654	820	515	256	308	4110
Net trade:							
Luxury	26	–6	–11	–36	42	–14	0
Basic	–18	–5	11	18	–17	11	0
Investment	65	10	–10	–13	–32	–19	0

Note: Amounts in US$ bn 1975.

Table 8.6(b): *South strengthens as crisis continues in North beyond 1990*

Economy	1	2	3	4	5	6	World
Output:							
Luxury	59	56	48	34	35	14	50
Basic	20	23	32	46	45	71	29
Investment	21	21	20	20	20	15	21
Factors:							
Skilled	38	44	21	26	35	22	33
Unskilled	41	36	31	43	37	53	39
Capital	21	20	48	31	28	26	28
Households:							
High-income	47	52	47	41	32	26	45
Low-income	53	48	53	59	68	74	55
Net trade:							
Luxury	1	0	–1	–8	21	–12	0
Basic	–1	–1	1	3	–6	2	0
Investment	5	2	–2	–5	–27	–15	0

Notes: Data are percentages.
Net exports are percentage of output by sector.

industrialised Group 6 countries, because of a lower rate of increase in the skilled labourforce in those countries. Thus, although the total income of low-income households increased, the number of low-income households increased even faster. This increased basic goods consumption, as seen by the shift in the composition of production in these economies. (By contrast, the redistribution towards high-income households and luxury production in the more industrialised economies of Group 1–3, in this and subsequent tables, reflects in part an increase in the size of the skilled population in these countries.)

At the international level, the *status quo* strategy had generally supported the interest of the dominant economies, although there were changes as the political influence of economies moved in line with their relative economic strength. Global inequality within and between nations continued to increase, and led to more frequent international and domestic conflict and repression. Many countries in the South were pushed towards economic, cultural, military and technological dependence and political instability. This inhibited balanced growth and the satisfaction of basic human needs in the South. Local crises became permanent, and there was growing dissatisfaction with existing models of development, whether 'Western' or 'Eastern'. Towards the end of this phase of development the growing crisis in the economies of the North, and the emergence of some economically-strong, newly-industrialised economies, provided the context for more independent action from, and interdependent development within, the South.

BREAKING THE MOULD: 1990–2000

At the beginning of this period, the increasing disunity within industrial countries enabled developing countries, collectively and bilaterally, to negotiate more effectively against the still dominant North. This was further facilitated by the increasing ability of the industrial Group 3 CMEA countries, relative to the market economies, to produce manufactured goods for many developing countries' needs. Some newly-industrialised economies had by now become effectively established as industrial powers, with a growing potential to compete with already industrialised nations in advanced, as well as traditional, sectors. But the continued development of this potential was increasingly frustrated by the protectionist policies of the North. Protectionism weakened the older links based on colonial and historical ties, and exchanges within the South grew. Some integration within the South took place vertically between countries with complementary 'endowments' (finance, technical know-how and abundant labour). Some took place horizontally between countries with common problems (in response to the need to develop tropical energy and agricultural resources, for example). International institutions based in the South were established to facilitate these exchanges, being funded in part through the surplus of the richer oil-exporting nations.

Although new links were established, the underlying principles governing much of the economic interaction *within* a more self-reliant South resembled those of the *status quo* scenario, and were subject to the same adverse tendencies. In other words, they represented a dilution of the radical strategy, in order to gain support from a sufficient number of Third World countries with different political orientations. These processes and domestic policy orientations are summarised in Table 8.7 and Table 8.8, while the outcome for economic growth and distribution in the year 2000 is summarised in Table 8.9, which shows the evolving pattern of international trade, the results of a partial slowdown in technical change, and protectionism in the North.

The partial failure of the strategy for the South reflected the lack of attention to structural differences and diversity of interests within and between Southern countries. In practice, it proved difficult to confront these potential sources of division while a simultaneously building collective front against a common problem — in this case, Northern intransigence. As has so often happened in the past, a unity which denies heterogeneity almost inevitably leads to the subordination of the weaker groups in the alliance to inappropriate guidelines. Intentionally or otherwise, the policies reflected more of the understanding and needs of the stronger parties. In addition, the costs and difficulties of delinking were underestimated in a situation of urgent change. These costs included reduced access to Northern finance, technology and markets, which, while they reinforced dependency overall, mainly benefited major actors in the South. These economic interests also created substantial political and economic obstacles to successful self-reliant activities.

For these reasons, once it became operational, the strategy increasingly operated largely in the interests of richer industrial and raw material producers in the South. The Group 4 newly-industrialised economies now produced capital goods, and the Group 5 oil-exporting economies supplied fertilisers and fuel for the rest, as well as luxury consumer goods. This created new patterns of technological

Table 8.7: *Dominant orientations and tendencies in collective self-reliance regime*

	Policy orientation	*Adverse tendencies*
Trade	Increase trade within South to maximum extent. Some trade with North needed to produce technology and essential produce technology and essential products but should operate on South's terms.	Problems of North–South trade and cooperation exist for South–South exchanges also.
Transnational enterprises	With greater independence from North, TNCs are more effectively controlled or nationalised.	TNCs recognise self-interest in protectionist South and manipulate transfer and trade between Southern countries.
Institutions	Southern-controlled regulatory bodies with linked free trade areas. Third World banks and research institutes. Strengthening of regional economic associations and political alliances.	Previous experience of economic and political cohesion in South is relatively poor. Internal cohesion in South weakened even without external threat of Northern imperialism. Multi-polar centres in South, often in conflict and competition.
Distribution, welfare and aid	South builds more egalitarian and less militaristic structures. Richer Southern nations assist the poorer. Attention paid to basic needs of low-income groups. Increasingly inegalitarian development in the North.	Increasing disparities within South continues. Insufficient attention paid to building up effective demand of low-income groups. Essentially relies on South-oriented market to provide 'trickle-down'.
Life-style	Southern cultural styles fostered with systematic divergences from North. In North increasing chaos and alienation in declining regions.	Pattern of development imposes dominant Southern styles in place of Northern.

231

Table 8.7: *Continued*

	Policy orientation	*Adverse tendencies*
Urban/rural	Essentially rapid industrialisation strategy. Some attempts to decentralise production. Scale of individual projects reduced.	Insufficient attention to rural development.
Orientation of production	With dependence reduced, production more oriented to domestic basic needs instead of export or import substitution.	Only marginal shifts towards production of basic goods occurs but with some tendency to produce more Southern-relevant products.
Technology	Much greater use of Southern productive and innovative skills, although some links to North inevitable. Move to improved traditional designs, many less sophisticated and smaller-scale. Set up Third World technology bank. Maximise technical cooperation in South for both appropriate and advanced technologies. Links to North designed to achieve maximum control of technologies and foreign firms. Setting-up of Southern controlled TNCs.	A high degree of delinking from North and development styles prove to be unfeasible. Present patterns of resources, finance, know-how and needs are not well-matched to needs. Insufficient consideration in strategy of structure of production, consumption and technology. Strategy is too piecemeal. Even transfer of techniques between LDCs remain under control of Northern MNCs or at best under South MNCs (ie. outside inter-governmental control).
Finance	Nationalise and limit foreign shareholding. Set up Third World development banks with favourable terms for cooperative projects and sanctions for stepping outside guidelines.	Surplus-rich Southern nations may not invest in other LDCs on terms consistent with CSR. Unless Northern demand kept up may be insufficient income to finance rapid industrialisation strategy.

232

Table 8.7: *Continued*

	Policy orientation	*Adverse tendencies*
Skills and education	Attempt to utilise local skills and entrepreneurs.	Model of R & D and educational institutions still largely based on Northern model. Management, innovation and marketing skills and experience limited.
Employment	Rapid growth with developing skills and demand for labour.	Rapid industrialisation strategy is unlikely to utilise all skills satisfactorily.
Raw materials	Producer-controlled stockpiles with quota agreements on sales to North. Greater efforts to use local and traditional raw materials. Differential pricing for poorer countries.	Some raw material exporters remain North oriented or require overly high prices from new Southern markets.

Table 8.8: *Collective self-reliance: scenarios for economic actors*

Group 1   Northern scenario is the result of failure to maintain *status quo*. Failure of world demand, Group 2 collapses and becomes increasingly protectionist and Group 4 and Group 6 delink, Group 2 retreats into isolationism. Increasingly competitive trade within Group 1, hence need to adopt policies of Group 2 in *status quo* scenario, i.e. worse income distribution and lower public spending and welfare benefits, high arms exports, lower rate of product, less investment, etc. especially in basic sector and hence lower technical change in basic sector. Impact of private sector bankruptcy in Group 2 or Group 4 passed to the public sector. Much more selective aid and investment in South and total level reduced.
Outcome: Technical change, increasing competition and declining world demand plus internal strife lead to progressive collapse in Northern economies, verging on military confrontation with Group 3.

Group 2   Tendency to become a peripheral economy of Group 1, e.g. TNC enclaves for luxury goods exports to Group 1. Consumption patterns still linked to Group 1, falls behind in technology, resulting in technological dependence on Group 1. More rapid collapse because of inability to implement wage restructuring to market level (i.e. comparable with Group 4) and modernise industry. Increasing competition from Group 1 and Group 3 with rapid increase in raw material costs from Group 5 leading to protectionism. Public basic expenditure falls. Arms expenditure rises, but arms exports fall. Tendency for domestic investments in non-productive industry or abroad — domestic rate of profit falls. Increase in unemployment, fall in effective demand, increased industrial strife, very low productivity, no domestic innovation. Exacerbated demise as South cancels debts. Little or no aid and investment to the South. BOP and budget deficit crisis.
Outcome: Protectionist policies fail to work as not backed up by integrated technology/distribution/economic structure policy.

Group 3   Economy cannot sustain domestic demand for luxury goods. External threat from Groups 1 and 2 (and vice versa) leads to increased arms spending. Increasing arms expenditure in LDC operations. Restrictions on technological exports from Group 1, no investment and export credits by Group 1. Declining export possibilities in North, hence hard currency purchases additionally difficult. Raw materials increasingly expensive. Some internal delinking in CMEA regions and Warsaw Pact, and increased costs of cohesion. Some trade links with South improve.
Outcome: Difficult growth, but begins to overtake Group 2. Non-military sectors of economy become increasingly backward. BOP deficit declines.

234

Table 8.8: *Continued*

Group 4	Scenario results from the bias of the *status quo* regime against the South and also a failure to establish a NIEO. Attempt to create a NIEO in the South by leading Group 4 and 5 countries. Rapid urban industrialisation strategy — investment and technological change concentrated in capital/luxury goods sector. Some delinking of luxury consumption and technology from Group 1 — reflects local factor prices — less labour displacing technical change although capital intensity does not decrease as fast as in the *status quo* scenario. Some income redistribution, but insufficient to create mass markets. Strong interaction with Group 6. *Outcome*: Initially benefical but demand does not expand sufficiently fast because of failure to redistribute domestically in South and collapse of export markets in North.
Group 5	Quota on raw material exports to North, concessionary rates to Group 4 and Group 6. Reduced demand and price for raw material exports but longer life income (i.e. costs of production of raw materials increase and surplus decreases). Becomes more dependent on Group 4 for technology, technology less sophisticated and unable to go to lower grade/tertiary extraction. *Outcome*: Relatively lower growth since income from South does not compensate for lost demand from North. Also difficulty transforming to industrial economies.
Group 6	Some direct investment from Group 5. Some preferential investment aid from Group 5. Becomes dependent on Group 4 for technology, although attempts to introduce more appropriate technology in basic sector. Eventually Group 4 and Group 5 set up enclave sectors in Group 6. *Outcome*: Tendency to become periphery states to Group 5 with somewhat higher income than in *status quo* scenario and better distribution and technology.
International institutions	Tension within Northern-dominated intergovernmental agencies. Partial devolution of Southern interest in established institutions and strengthening of new Third World secretariat and Development Bank. Political alliances to control and facilitate exploitation of common resources (e.g. ocean outer-space). Renewed emphasis on regional economic development pacts, communities and planning. Regulation of Northern transnationals, nationalisation and joint ownership and tighter licensing agreements. Many Northern TNC subsidiaries divest into the South. In the North TNCs increasingly dominate national and international economy.

Table 8.9(a): *Summary social accounts for all economies: world economy at start of new international order of year 2000*

Economy	1	2	3	4	5	6	World
Output:							
Luxury	3883	1486	1574	668	248	145	8004
Basic	1267	575	974	799	338	732	4686
Investment	1396	535	609	349	155	167	3211
Factors:							
Skilled	1368	715	398	352	188	127	3149
Unskilled	1179	487	476	448	167	246	3004
Capital	592	230	729	333	120	105	2109
Households:							
High-income	1675	808	852	455	139	119	4048
Low-income	1415	629	763	699	330	377	4213
Net trade:							
Luxury	20	–5	–13	–33	46	–16	0
Basic	–15	–5	10	18	–19	11	0
Investment	45	5	–7	–6	–22	–14	0

Note: Amounts US$ bn 1975.

Table 8.9(b): *World economy at start of new international order of year 2000*

Economy	1	2	3	4	5	6	World
Output:							
Luxury	59	57	50	37	33	14	50
Basic	19	22	31	44	46	70	29
Investment	21	21	19	19	21	16	20
Factors:							
Skilled	44	50	25	31	40	27	38
Unskilled	38	34	30	40	35	52	36
Capital	19	16	45	29	25	22	26
Households:							
High-income	54	56	53	39	30	24	49
Low-income	46	44	47	61	70	76	51
Net trade:							
Luxury	1	0	–1	–5	19	–11	0
Basic	–1	–1	1	2	–6	2	0
Investment	3	1	–1	–2	–15	–9	0

Notes: Data are percentages.
Net exports are percentage of output by sector.

dependency. Vertical intergration began to rigidify into new economic hierarchies within the South, while horizontal integration stimulated national competition as much as regional cooperation. Poorer developing countries effectively transfered their dependence from the North to the South. But the South did insulate itself, to an extent, from the repercussions of the problems in the North, and largely avoided attempts by the North to frustrate Southern unity.

A major problem of this strategy came from the failure to create mass consumption markets in much of the South, and the associated persistence of large-scale poverty in the poorer countries (and the larger part of Africa). Increased intra-South trade provided a temporary boost to partially offset the decline in exports to the North. But there was a general failure to redistribute income and stimulate the necessary domestic demand in the South. Thus the human and ecological crises in the poorest countries continued to deepen.

THE EVOLUTION OF THE NEW INTERNATIONAL ECONOMIC ORDER, 2000–10

By the end of the century, the inevitable catastrophe that would result from allowing international tensions to grow further was recognised by social movements and political institutions around the world. A concerted effort by a significant group of nations to bring about a NIEO emerged from the situation of deepening crisis depicted above. This change depended upon major political change away from conservative strategies in at least a number of leading Western countries, and a parallel liberalisation in Group 3 countries. The growth of a new middle class in many Third World countries also provided internal sources of change in these regions. Governments were under pressure to find real solutions. New leaders emerged. And a series of major international summit meetings and popular conferences and workshops began to establish a framework of mutual trust, and laid the basis for new 'bridge-building' institutions.

The reasons for this step forwards — or step back from the brink — were varied. They derived both from initiatives of the rising Southern institutions and, in the North, from public reactions to the threat of nuclear holocaust, the collapse of work under the impact of new technologies, and deteriorating social provision. For the North, self-interest determined a willingness to make substantial concessions to the South, while the Southern countries realised a need for expanded export markets and the usefulness of some new technologies and

Table 8.10: *Dominant policy orientation and tendencies in the new international economic order regime*

	Policy orientation	*Adverse tendencies*
Distribution, welfare and aid	Attempt to reduce poverty in South through aid. Redistribution and welfare policies in North.	More aid tends to increase dependence and worsen some income distribution in the South.
Life-style	Greater diversities of life-styles within and between regions tolerated. Growth of informal economies in North.	Diversity tends to function as a means of product differentiation and establishing superficial cultural identity. Still inadequate mass consumption in South.
Urban/rural	In LDCs industrial growth still dominates with limited additional support to traditional agriculture. Some attempts to prevent rural decline.	No emphasis in South on rural economy except for limited development of agriculture technology.
Technology	Innovation in LDCs encouraged and directed through government agencies. More rapid transfer of sophisticated techniques and regulations to encourage TNCs to release technical knowhow. R & D encourage for appropriate techniques and joint patenting.	Most competitive techniques are not transferred or are controlled by TNCs or advanced economies. Appropriate techniques are either a palliative in non-profitable low-income sectors or patents are appropriated by TNCs. Transfer between developing countries still controlled by TNCs.
Finance	Marshall Aid-type plan for Third World and investment at favourable interest rates. Greater local participation and joint ventures controlled by South. Savings from arms spending used for development. Insurance provision for Third World entrepreneurs and small industries.	Investment (and aid) coupled to terms which direct development in interests of donors, and politically oriented between major power blocs and dependent nations.

Table 8.10: *Continued*

	Policy orientation	Adverse tendencies
Skills and education	Increased training and educational aid by North for South. Greater innovative skills seen in South, particularly in traditional/basic sectors.	Southern innovative skills still under-utilised and training oriented to Northern-oriented production and consumption.
Raw materials	Improved terms of trade. Joint control of buffer stocks and long-term agreements on prices and levels of production.	Increased prices offset by attempts of North to conserve resources, terms of trade are constantly changed.
Transnational enterprises	Some degree of international surveillance. TNCs operate within new self-interested logic, which is increasingly differentiated from interests of North countries.	This new logic is not necessarily in the interests of the South and TNCs still manage to bypass controls.
Orientation of production	Policies to stimulate effective demand for Southern products. Some marketing agreements to prevent entry of unsuitable products to South.	Still largely imitative and oriented to dependence. Transfer of production worsens income distribution in North. Transer of production unlinked from transfer of technology may worsen net gain to South.
Employment	Full employment or creative occupations desired. Greater support for small industries in North and South.	Policies too limited to make a significant impact on unemployment. Increasing tendency for 'black' economy to develop in less advanced Northern economies.
Trade	Import controls to protect 'infant' industry in LDCs to North and improved in terms of trade for South raw materials and manufactures.	Less effective until it forms part of package to change production and consumption domestically.
International Institutions	Strengthening of international agencies, including UN Central Bank. Growth of regional economic communities.	Institutions are more aligned to South interests but power still remains unrepresentatively with Northern interests.

Table 8.11: *New international economic order: scenario for economic actors*

Group 1	Similar to *status quo* with following changes: lower arms spending; greater aid *(pro rata* to p.c income) and foreign investment on favourable terms. Some foreign debts cancelled some small domestic redistribution; higher raw material costs/resource saving higher; reduced arms exports, fewer barriers to imports from Group 3 to Group 6. Slower technical change on aggregate than in *status quo* for Groups 2–6 leads to lower investment. New investment and technical change in basic sector is higher because of redistribution and higher aggregate demand. Relatively higher growth of capital goods industry for export. Greater rate of transfer of technology and a narrower gap between Group 1 and the technology used in other economies, especially the luxury and capital goods sectors. *Outcome*: Relatively successful with increased BOP surplus.
Group 2	Similar to *status quo*. On balance, Group 2 lags slightly less far behind Group 1 than in *status quo* scenario. Group 2 pay proportionately less in aid than Group 1. Welfare payments for unemployed higher than *status quo*. More rapid penetration in basic and luxury sector from Group 3. Group 4 and Group 1 get most key contracts from Group 3–5 for capital goods and (reduced) arms exports to South. *Outcome*: Still experiences relative slowdown in growth and gains relatively less than Group 1 from the NIEO.
Group 3	Reduced arms spending leads to very significant increase in productive investment. Investment aid internationalised from Groups 3–6. Narrower technological gap between Group 3 and Group 1. Increase in domestic production and imports of luxury goods. Increased imports of sophisticated capital goods and joint ventures with Group 1. Increase in BOP deficit. *Outcome*: Overall higher growth than in *status quo* scenario, especially because of arms spending reductions, increased transfer of technology and relatively protected markets.
Group 4	South still largely imitative of North. Greater exports to North stimulates luxury sector, urban industrialisation-oriented strategy, export-led. Greater transfer of technology gives more rapid technological change in capital and luxury sectors. Once again a higher demand for import substitution luxury goods, and increased import of sophisticated capital goods which cannot be manufactured domestically. Reduced arms imports, spending and domestic production. Higher domestic investment and locally retained surplus. *Outcome*: Overall beneficial to Group 4 because of increased investments, favourable transfer of technology and reduced arms expenditure. Increasingly rapid overtaking of Group 2.

Table 8.11: *Continued*

Group 5	Export prices for oil index-linked. Increased prices for exports and increased demand partly offset by resource-saving policies in Group 1 and Group 2. Some concessional prices to Group 6. Greater imports of all types of good – but have higher luxury component than *status quo* scenario. Consumption basket and some technology rapidly approach Group 1 and overtake Group 4. Increased foreign investment to Groups 1, 4 and 5. Reduced arms imports and spending. *Outcome*: More rapid growth than in *status quo* scenario or CSR although structural and distributional problems are not reduced.
Group 6	Greater foreign investment and crisis aid initially directed to low-income group, nominally according to *per capita* income from Groups 1, 2, 3 and 5. Technological sophistication in luxury goods sector, some growth of capital sector – joint ventures rather than TNC enclaves. Some capital-saving techniques in basic sector (appropriate technology). Some attempt to redistribute income. More favourable terms for exported basic goods. *Outcome*: Some improvement over earlier strategies but still impoverishing growth.
International institutions	Revival of old, and setting-up of new, inter-governmental institutions. Southern views more effectively represented in new system much is oriented to the regulation of distribution of the proceeds of development. Increasing regulation of the private sector and more integration between Northern and Southern TNCs, with joint ventures at governmental and regional level. Controls on international finance, etc. International income tax based on *per capita* income and worldwide earning with redistribution back to LDC producers.

skills developed in the North. The underlying approach to policy, and the processes underway at the domestic and international level, are shown in Tables 8.10 and 8.11.

The consequent reduction in armaments spending had a significant impact on the relative growth rates of national economies. The funds could now be invested in productive areas of the economy, and used to support some restructuring in the international economy and development projects in the South. Research efforts could be diverted away from destructive goals to a range of human and social needs. Greater control of the international economy was restored to governments and inter-governmental agencies via regulation of trade, aid and technology. There was also some shift in economic power in the private sector from transnational to national actors, through nationalisation and the setting-up of joint ventures between governments and transnational enterprises. Regulation of the level of international transfers included the suspension and, in some cases, the cancelling of still outstanding international debts for many developing countries. This was sufficient to bring about greater stability in the world economy, and to support shifts towards a more equitable distribution in line with the objective of a revitalised United Nations system.

Many countries experienced increased economic growth, and significant redistributive policies were effected as depicted in Table 8.12. However, even by the year 2010, for many of the poorer nations the strategy gave only marginal respite. Further, the new international expansion was threatened by instability arising from inflationary pressures caused by high levels of government and intergovernmental expenditure, and the renewed demands from a range of sectional interests for a larger share of the growing income.

New threats to the strategy became apparent as the crisis diminished, and the self-interest of nations was increasingly visible in their negotiating stance within the new institutions. In many respects, the strategy was still over-dependent on the goodwill of the North towards the South. The North did not always meet development aid targets and other obligations reliably – for example, those involving the sharing of scientific and technological advances, or regulation of the use of ocean and space resources. East–West ideological conflict also re-emerged even as Group 3 countries continued to diverge. The Western powers were again responsible for increased arms expenditure. Partly this may have been a way of diverting Group 3 expenditures by engaging these countries in an arms race; more obviously, it reflected the fact that arms exports to the Third World were still a most profitable area of trade, while arms production

Table 8.12(a): *Summary social accounts for all economies: NIEO survives as poorest economies restructure after year 2010*

Economy	1	2	3	4	5	6	World
Output:							
Luxury	4971	1714	1939	1054	337	199	10213
Basic	1489	665	1153	1069	494	1002	5872
Investment	1903	634	758	579	265	277	4416
Factors:							
Skilled	2190	956	581	613	318	236	4894
Unskilled	1072	439	440	520	169	289	2928
Capital	758	267	935	544	216	148	2868
Households:							
High-income	1786	846	975	706	193	164	4671
Low-income	2104	806	1001	1031	519	559	6019
Net trade:							
Luxury	81	−7	−24	−74	61	−37	0
Basic	−37	0	17	26	−20	14	0
Investment	86	17	−13	−12	−51	−27	0

NOTE: Amounts in US$ bn 1975.

Table 8.12(b): *NIEO survives as poorest economies restructure after year 2010*

Economy	1	2	3	4	5	6	World
Output:							
Luxury	59	57	50	39	31	13	50
Basic	18	22	30	40	45	68	29
Investment	23	21	20	21	24	19	22
Factors:							
Skilled	54	58	30	37	45	35	46
Unskilled	27	26	22	31	24	43	27
Capital	19	16	48	32	31	22	27
Households:							
High-income	46	51	49	41	27	23	44
Low-income	54	49	51	59	73	77	56
Net trade:							
Luxury	2	0	−1	−7	18	−19	0
Basic	−2	0	1	2	−4	1	0
Investment	5	3	−2	−2	−19	−10	0

Notes: Data are percentages.
Net exports are percentage of output by sector.

243

was a means of shoring up high-technology industries challenged by Group 4 competition.

In this period the new international institutions were better able to accommodate the more serious threats to the newly-emerging world order than previously. Faced with problems similar to those of the 1970s, the international institutions were again confronted with a major challenge to their authority and effectiveness. This challenge was contained largely because the successive traumatic experiences of the twentieth century had forced the widespread realisation that some compromises must be made if global tensions were to be successfully defused. In this respect, a positive global learning process had taken place. The strain produced by the restructuring of the world economy was cushioned by compensatory transfers around globally-negotiated objectives. Thus, despite a continuing tendency for policies to operate in favour of dominant nations in the North and South, problems of inflation and conflict were kept in check by an increasing shift of national autonomy to regional and global international regulatory agencies.

While the NIEO period generally provided an improved outcome for poorer nations, and social groups, compared to the *status quo* scenario, for the most part these nations were unable to achieve sustained industrial development. Without substantial amounts of development aid, they were unable to provide even minimal satisfaction of basic human needs for the majority of their populations. Levels of aid were higher than in the past, but so too were the problems they confronted. Like its predecessors, the reliance on strategies of export-led growth and provision of emergency relief proved inadequate for considerable numbers of Group 6 countries. But the volume of exports that these countries required to achieve successful export-led development still far exceeded the richer countries' capacity (or willingness) to import their goods. Thus, many of the poorest countries were obliged to seek new development paths.

HUMAN NEEDS DEVELOPMENT: 2010 ONWARDS

Unlike earlier periods, this phase did not evolve as a global development strategy, although the cooperation of the international community was an important component of its success. Rather, a new programme was undertaken by individual less industrialised countries, and small groups of these countries, while the rest of the world continued to pursue a NIEO. It arose from the failure of other

Table 8.13: *Dominant policy orientation and tendencies in the human needs-oriented regime*

	Policy orientation	*Adverse tendencies*
Distribution, welfare and aid	Long-term aim to create a production structure which is more egalitarian and is robust against political interference. Income redistribution alone is likely to be ineffective, and is combined with structural technical and social change. Income transfers needed at international and domestic level in the short-run.	Internal and external political opposition to social change threatens reforms.
Life-styles	Ideal is to have decentralisation of national control. Greater participation in social and workplace decision-making. Attempt to maintain diversity and its appreciation as a counter to centralisation via other institutional reforms.	Strong institutional reforms bought at some cost of personal and social freedoms for significant part of population.
Orientation of production	Production oriented to basic consumption goods and capital equipment and to minimise dependence on outside.	For smaller countries a high level of autonomy is not possible or involves high prices of many goods.
Urban/rural	Greater emphasis on development of rural sector. As agricultural surplus increases this is used to promote rural industry, rather than urban growth.	Improved quality of social and economic life cannot be provided by small-scale social units only. Decentralised production and consumption and social organisations are inefficient.

Table 8.13: *Continued*

	Policy orientation	Adverse tendencies
Technology	Initially much greater use of traditional techniques and skills which would develop to absorb scientific inputs. Aim is for distinctive technologies and products to evolve from this base. Sources of technology and institutions should be such as to enable LDCs to absorb accumulated knowledge of North but to apply it to own problems. Technologies should be transferred earlier in the product cycle so that both product and process can be developed towards local needs and resources.	Low-level techniques have limited possibilities. LDCs do not develop own techniques sufficiently rapidly with risk that they become permanently delinked from North. Labour-intensive techniques demand managements and marketing skills which LDCs do not have.
Finance	Diversion of luxury consumption and arms expenditure into basic consumption provides major source of new investment needs. All foreign joint ventures to have strict control. Finance and risk insurance for new ventures to be made available to peasants and small enterprises. External aid provided via UN Central Bank at concessionary rates.	Cost of organising and servicing small-scale enterprises is high and institutionally very difficult. International agencies, foreign investors and local firms see strategy as risky.
Skills and education	Education oriented around local needs. (At present skills which are used are defined by and oriented towards interests of dominant countries and groups.) Distinctive elements of each culture embodied in technology as it evolves.	Many LDC skills are obsolete and cannot bridge the gap. Lack of innovative tradition at indigenous level with risk that LDCs are left technologically backwards.

Table 8.13: *Continued*

	Policy orientation	*Adverse tendencies*
Transnational enterprises	Ensures TNCs support national interests. Only ventures with majority national control acceptable with full training, etc. and transfer of control.	Too many options make avoidance of regulations prevalent.
Employment	Full employment is a major goal of strategy with participation in formal sector designed to strengthen political base of presently deprived groups. Patterns of employment more flexible.	Egalitarian and participative dimension of work structures tend to be unstable and inefficient, and new power structures emerge to challenge more equalitarian goals of society.
Trade	External links are oriented to domestic requirements of the human needs strategy. Most advantageous trade policy depends on economic types, sectors and time-periods.	Whenever trade needs of LDCs run counter to those of richer trading partners they are threatened.
Raw materials	Some countries able to use raw materials to underwrite strategy. International agreements set long-term prices and quotas. Locally available materials used domestically as far as possible.	Tendencies for international agreements to fail or require constant revision. General principle of using local resources (including skills) leads to inefficient use of innovative capability.
International institutions	International agencies restructured to ensure that needs of LDCs are more systematically met.	Attempts to undermine international framework, especially by less successful richer economies.

strategies, including the NIEO, to ensure their sustained development. For some of these poorer countries, an initial boost for more independent development was gained from development aid transfers, for others from the exploitation of raw materials, but within a modified NIEO with a framework of agreed long-term price structures. This represented a considerable shift in policy: the new donor economies of North and South were now prepared to recognise that, before the poorer countries of the South could enter the world system without exacerbating their social problems, they must acquire the technical and other capabilities needed to secure social objectives. Thus, the donors' relatively minor economic interests in Group 6 countries were offset by longer-term interests in the development of the poorer economies as part of a more humanitarian vision, and a concern with the instability that might result from continuing inequalities of this magnitude.

In contrast to earlier strategies, the principal element of a human needs strategy in poorer countries (shown in Table 8.13) centred developments on the skills and consumption needs of a large, low-income population, with an emphasis on rural rather than urban development. This entailed significant domestic political reforms and a restructuring of the international economic relationships of these countries, to prevent the process of internal development being undermined by external links. Because of this, the strategy operated against the interests of many entrenched international and domestic interests, and there were efforts to destabilise it. The strength of links between interested parties in the North and South were sometimes effective in preventing cooperation between developing countries. Consequently, those nations which could adopt the strategy most effectively, were often obliged to do so within a framework of national self-reliance, and rely on bilateral relationships with more industrialised economies in the North and South rather than undertaking a full programme of collective self-reliance.

The support available internationally, however, allowed these developing economies to take selective advantage of the benefits of international trade and technological change. But against this, the scale of social and environmental problems in these poor countries was so large that the greater satisfaction of the masses' material needs was in many cases only achieved with a considerable curtailment of political liberties. Only beyond the year 2020 has this situation begun to change. Data for the year 2020, (Table 8.14) show that some progress has been made, but the long-term future is still uncertain. There are many problems yet to be overcome as these

Table 8.14(a): *Summary social accounts for all economies: NIEO and human needs-oriented development continues past year 2020*

Economy	1	2	3	4	5	6	World
Output:							
Luxury	6167	2105	2511	1810	525	288	13406
Basic	1777	761	1402	1461	664	1955	8020
Investment	2717	807	1019	961	513	371	6388
Factors:							
Skilled	2759	1151	677	976	528	387	6478
Unskilled	1314	549	568	696	227	577	3932
Capital	1072	326	1257	933	335	241	4165
Households:							
High-income	2357	938	1028	1014	238	152	5726
Low-income	2632	1075	1464	1621	892	1167	8850
Net trade:							
Luxury	70	–16	–33	–71	64	–13	0
Basic	–34	13	21	38	–63	25	0
Investment	120	16	24	4	–39	–125	0

Note: Amounts in US$ bn 1975.

Table 8.14(b): *NIEO and human needs-oriented development continues past year 2020*

Economy	1	2	3	4	5	6	World
Output:							
Luxury	58	57	51	43	31	11	48
Basic	17	21	28	35	39	75	29
Investment	25	22	21	23	30	14	23
Factors:							
Skilled	54	57	27	37	48	32	44
Unskilled	26	27	23	27	21	48	27
Capital	21	16	50	36	31	20	29
Households:							
High-income	47	47	41	38	21	12	39
Low-income	53	53	59	62	79	88	61
Net trade:							
Luxury	1	–1	–1	–4	12	–4	0
Basic	–2	2	1	3	–9	1	0
Investment	4	2	2	0	–8	–34	0

Notes: Data are percentages.
Net exports are percentage of output by sector.

Table 8.15: Household income and distribution from 1975–2020

Economy group	1	2	3	4	5	6
Year 1975 —	*Base year of future history*					
High	11.0	10.9	8.7	6.0	7.5	0.87
Low	3.9	3.3	2.1	0.99	1.3	0.09
Low/high	0.35	0.30	0.24	0.16	0.17	0.11
Year 1980 —	*Worsening worldwide economic crisis*					
High	11.4	11.2	9.6	6.1	6.7	0.84
Low	4.4	3.8	2.6	1.1	1.4	0.10
Low/high	0.42	0.35	0.27	0.19	0.21	0.12
Year 1990 —	*South strengthens as crisis in North continues*					
High	15.3	14.3	11.8	7.3	8.6	0.81
Low	4.4	2.7	2.4	1.0	1.2	0.10
Low/high	0.29	0.20	0.20	0.14	0.14	0.12
Year 2000 —	*Start of new international economic order*					
High	18.1	15.0	13.1	8.0	7.7	0.75
Low	3.8	2.6	2.1	1.2	1.2	0.10
Low/high	0.21	0.17	0.16	0.15	0.15	0.13
Year 2010 —	*NIEO survives as human needs strategy begins*					
High	18.8	14.9	14.4	11.0	9.0	0.96
Low	5.5	3.1	2.6	1.6	1.5	0.14
Low/high	0.29	0.21	0.18	0.14	0.17	0.14
Year 2020 —	*NIEO and human needs strategy continue*					
High	24.3	16.1	14.9	15.1	10.2	0.84
Low	6.3	4.0	3.7	2.4	2.4	0.27
Low/high	0.26	0.25	0.25	0.16	0.24	0.32

Note: Amounts are *per capita* income in US$ 000s (1975)..

countries now begin to reintegrate into the world economy, and world society – hopefully – achieves an epoch of global mutual development.

THE OUTLOOK FOR DISTRIBUTION

What are the prospects for *per capita* income and distribution in this future history? Table 8.15 shows how the *per capita* income for the

Table 8.16: *NIEO and human needs-oriented development continues past year 2020: Social accounts for least industrialised economies*

	Production sectors			Factors of production			Final demand			Total
	Luxury	Basic	Invest.	Skilled	Unskill.	Capital	High-inc.	Low-inc.	Foreign	
Luxury	15	147	21	0	0	0	10	108	-13	288
Basic	76	782	85	0	0	0	70	917	25	1955
Investment	28	217	37	0	0	0	71	142	-125	371
Skilled	60	157	171	0	0	0	0	0	0	387
Unskilled	69	484	23	0	0	0	0	0	0	577
Capital	39	167	35	0	0	0	0	0	0	241
High-income	0	0	0	127	0	69	0	0	43	239
Low-income	0	0	0	260	577	172	87	0	70	1167
Total	288	1955	371	387	577	241	239	1167	0	5224

Note: Amounts in US$ bn 1975. Rows and columns may not total precisely to figures indicated due to rounding.

251

household groups described in the model shifts from 1975 to 2020. Typically, in each economy average real incomes have doubled. *Per capita* growth has been most rapid in the newly-industrialising countries (but averaging less than 2 per cent), and slowest for the less dynamic industrial nations (averaging slightly over 0.5 per cent). For the poorest countries the *per capita* growth rate has been around 1.5 per cent but this is largely concentrated in the last phase of human needs-oriented development. The irregular pattern of growth from period to period follows from the assumption made. The changes in distribution, to, follow from assumptions about technical change, the restructuring of domestic production and international trade, and domestic and international transfers.

With respect to distribution within economies, there are some features in common. Distribution between households improves only in the initial period but worsens thereafter until the year 2000, or even beyond. The initial improvement in distribution comes from a decline in the profit share for each economy, which induces a period of technical change. Up to 1990 significant amounts of unskilled labour are displaced. With falling overall growth and increasing unemployment among unskilled workers, the distribution in most regions worsens.

After 1990 in the South, and after the year 2000 in the North, with increased regional growth the low-income households benefit from the growing income to skilled labour, and the more welfare-oriented policies which are implemented domestically as well as internationally as part of the NIEO. The departures from this broad picture are evident in the slower growth of the Group 2 economies, and the somewhat less marked and delayed shifts in distribution of the Group 3 economies. Group 4 economies, as noted above, exhibit the highest growth rates and the most stable distribution, as their demand for labour is comparatively strong. By contrast, the growth of the oil-exporting economies is slower. This is mainly because the overall slow rates of growth assumed for the major industrial oil importers are not offset by new demand from the South; also, demographic growth in these countries is somewhat higher than in the newly-industrialising countries. Incomes in the poorest countries remain especially low, although income distribution improves substantially in the last period.

Over and above these trends, the results show some unevenness. Distribution is especially sensitive to the assumptions concerning population growth and technical change, as well as to more direct redistributive mechanisms (such as tax-based transfers and development aid). It is possible to select parameter values which smooth out this unevenness; however, given that parameters represent events which are to some extent independent of each other, we should expect to see some fluctuations beyond those implied by the scenario.

All these forces contribute to the final pattern of distribution, and this is again illustrated with reference to the Social Accounting Matrix for the least industrialised Group 6 economies, shown in Table 8.16. Cumulative changes have been made to the factor payments from each sector, in line with the experiments described in Chapter 7. One point should be noted, however: in the last decade covered by the future history, overall economic growth in the economy is high compared to the 1970s, while demographic growth has substantially declined. Consequently, the shift to low-skill intensity in technology is rather less than suggested by our earlier calculations for appropriate technology. Thus, the technology we have asumed for this last period is conditioned by the growth of distributional requirements of the human needs-oriented strategy at a particular point in history.

The SAM shows that by the year 2020 there is a large shift in the distribution of investment income and foreign aid towards low-income households. By this time, also, skilled labour is contributing more than two-thirds of wage income to low-income households. Consumption and output both comprise a greater share of basic goods, and the economy becomes overall more self-sufficient. (The exception to this is the considerable volume of capital goods whose purchase is covered by the increased development aid.)

Without the substantial shifts in economic policy assumed for the period after the year 2010, the outlook for income distribution in our future history is dismal. The main considerations in developing the scenario have been political and institutional, and these have governed the assumptions concerning growth rates, technical change, domestic and international transfers, and so on, which ultimately determine the pattern of income distribution revealed by the model. By relaxing the constraints imposed through institutions, it would be possible to construct a scenario in which the real incomes of the poorest group increase at a steady 5 per cent over the entire period covered by the scenario. This would bring them up to the level of the richest income groups in 1975, by the year 2020. Given unchanged demographic data, this would demand that the total income of the poorest economies rise by some 6 to 8 per cent each year. For such a sustained growth rate to be feasible requires changes in national and global institutions far more sanguine than those contemplated here. Only after the initiation of a human needs-oriented strategy for the least industrialised countries, fostered at a global level, do growth rates of this magnitude become compatible with the institutional changes envisaged. And even then, with the growth rates assumed in the last period, the incomes of the poorest households considered could match those of the highest until far into the twenty-second century.

Despite the belated positive policies, and despite the continued crisis in the economies of the North, the gap between rich and poor across the world remains enormous. In the base year 1975, the ratio of these incomes was almost 120:1. By the year 2000, this worsens to 180:1 and even in the year 2020, after two decades of the new international order and one dcade of human needs-oriented development, the ratio has fallen only to 90:1. Few people can view such a prospect with equanimity.

POSTSCRIPT: TOWARDS GLOBAL MUTUAL DEVELOPMENT

We have presented our future history as a sequence of policies, implemented with different degrees of success. At the outset, it was emphasised that this was only one of many possibilities. Alternatives may be considered more likely than that given here. What we have done is to illustrate the evolution of a number of distinct strategies, and to capture something of the dramatic changes that are likely to lie ahead. Thus, we have not outlined a *prediction*, but illuminated the range of possibilities more fully.

It is customary to label future studies as either 'optimistic' or 'pessimistic'. Our view is that the number and magnitude of crises in the present era are sufficient to defeat facile optimism about the future. Some readers may accuse us of cynicism, others of wishful thinking in our descriptions of the nature of global and domestic political institutions. But the future described is ultimately far more optimistic than many others which could be presented, and which are the substance of much political discourse today. At each of the major 'branching points' covered by the scenario, a relatively positive step is taken; and the strategies pursued are assumed to partially further the interests of at least, their principal advocates.

Above all, the scenario is optimistic in that it avoids the terrible consequences of a global nuclear war. We do not ascribe the growing danger here simply to the implementation of conservative economic policies, since many other forces — for example, political intransigence and nationalism — play their part, too. With the assumed continued crisis among the Northern economies, the global threat of nuclear war is increased. Even though the North suffers no direct military engagement, surrogate wars between the superpowers may well be enacted in the territories of the South, thus crippling development there. We have, again, been optimistic in assuming this not to be the case, on any scale.

To have assumed that the world crisis will end during the 1980s would provide insufficient time for the necessary strong institutions in the South to become established. The creation of substantial Third World institutions has already begun, but in very few areas are these able to challenge the dominance of those representing the North effectively. While an earlier end to the present economic crisis might, in the short run, increase the overall rate of growth of developing economies, such change would merely be a continuation of past 'trickle-down' policies. An end to the crisis might, therefore, increase the income of some developing countries and provide opportunities for some newly-industrialising economies, and groups of raw material producers to emulate the oil-exporting economies of the 1970s. But, on balance, this is more likely to strengthen the position of the stronger economies of the North, which, with rapidly changing technology, could regain their comparative advantage for production of industrial goods. The gap between rich and poor economies would continue to widen and the international *status quo* substantially would remain.

Our scenario analysis portrays attempts to implement policies that are largely in the self-interest of restricted national and social groups as contributing to global tensions. The transition from the brink of world war to a systematic and widespread effort to rebuild new international institutions depends upon the emergence of strong movements for peace and international development, movements that are aware of the contradictions of existing development paths, and who provide a basis for more farsighted world leaders to gain power. Our scenario is optimistic in this respect, and in supposing that the new institutions which do emerge will be both strong enough and flexible enough to deal with the many evident conflicts which must emerge. Perhaps we have been pessimistic about the speed with which such transformations may be achieved: however, we have to admit that there are deeply-rooted obstacles facing these social movements despite the widespread unease and dissatisfaction with contemporary political institutions and ways of life.

The scenario is also optimistic about the success of a more inward-looking and self-reliant approach to development by the countries of the South. Unlike some of the political formulations of this approach, we recognise that it will confront considerable internal and external pressures seeking to undermine it. We have assumed that new institutions can be evolved by Southern countries to give them an influential role in effective global negotiations. Again, this means an optimistic view of the emergence of independent political forces in these countries: and while their history of struggle certainly leads us to expect continual re-emergence of repressed movements, it is by no means evident that they will escape the pitfalls of nationalist and authoritarian ideologies.

We have been less optimistic about whether many of the least developed nations, and the lowest income groups, will be able to ensure that their needs are respected in any of the global strategies. Ultimately, this requires empowering the poor — or reducing the strength of entrenched interests which oppose these needs. These countries and people need to escape from their role as pawns in a global power struggle, with little chance of gaining support for domestic programmes outside of military and ideological alliances. This requires East–West détente, and the limitation of new sub-imperialisms. The restructuring of the world political economy envisaged at the turn of the century provides, in our scenario, the preconditions for a human needs-oriented strategy to function, and for the countries implementing it eventually to reintegrate into the world economy. In this respect, the timescale for the sequence of events to take place may be too short.

Not to have assumed such breakthroughs in international and domestic policy would have been to accept that the historical process endlessly repeats itself — or worse. Obviously, the details of this future history could have been very different — but could they have been better? For the mass of the population in the poorest nations, much more could have been achieved — a doubling of real income for a family in the conditions found in many parts of the world today represents a pathetic improvement. The technical capabilities and material resources of the world are such as to give us the possibility of abolishing absolute poverty, as demonstrated in our own *World Futures: The Great Debate* (Freeman and Jahoda, 1978) and many other studies. By contrast, the scenario we present is bleak, but it serves as a negative example. It demonstrates the importance of moving much more swiftly towards policies which can ensure a sustained improvement in the situation of the poorest nations comprising roughly one half of the world's population.

Assuming that an earlier start towards a more desirable global path of mutual development was possible, then what have we learned from our study? Human needs-oriented strategies in the least industrialised economies, require positive assistance from wealthier countries. The redistributive aspect of a human needs strategy in the early steps to industrialisation would be facilitated by necessary financial and other inputs from outside. Ensuring that new technologies and know-how are available to these economies, as they move from poverty-reducing strategies to sustained development, is essential. We are less hopeful about the success of extreme autarkic policies but, even so, they may be preferable to externally-imposed policies which systematically undermine local efforts to bring about more just societies.

In this respect, the most obvious contradiction exposed by the experiments with the model is that when industrial countries offer development finance only on condition that other policy instruments (such as regulation of international trade) are not employed, the gift is effectively like giving water in a sieve. Whatever the extent of crisis in the rich industrial countries, the most acute and desperate problems remain with the people of the least developed economies. To achieve the social objectives which are at the heart of proposals such as those of the Brandt Commission, a suitable range and real choice of policy options must be available to developing countries. Of fundamental importance is the option to choose how domestic and international markets are to be used in the interests of their own economies. If the international market is taken as the ultimate arbiter of social policy, developing countries are likely to forgo their key development objectives. It is one matter for development policy to make use of the 'magic of the market', but quite another matter for it to be bewitched by it.

The experiments show that there is not necessarily a common self-interest for the North, as is suggested by reformist proposals. Different economic groups within the North gain or lose relative to each other from joint development with the South. This divergence of interests is likely to delay the implementation of these recommendations, unless it can be accommodated by recognising a more global community of interest. For the present, it seems imperative that the North reorients economic recovery around long-term global needs, rather than responding only to short-term national pressures.

Although we have expressed a number of specific reservations about the model used to examine the issue of income distribution, it does demonstrate vividly that poverty and income distribution are truly global issues. In an interdependent world, changes in one region can have a very significant effect on the consequences of social policy elsewhere. Even if we cannot always put much faith in the size (or even the direction) of the changes, the model demonstrates the importance of considering the interrelations in a more direct manner than is usually done in futures studies and policy research. The conventional wisdom of economics is based on very simple models and controversial empirical findings. But, as noted in Chapter 6, when economists include more actors or more relationships in the simple models, received wisdom is increasingly disputed — even before a rethinking of worldviews takes place.

We have relied on conventional theories for our analysis, but we could equally have argued in our scenarios that the depth of present or future crises will be sufficient to lead to new schools of thought. After all, our circumstances of crisis and far-reaching change are comparable to those faced by Smith, Marx or Keynes. At the very least, the assumptions in the present standard economic models against which 'paradoxes' are judged, must surely be revised. The analysis in this book has utilised different theories of development and techniques, but not in an attempt to create a new, unified theory. Rather we have tried to create a method with which to follow the logic of a world in which different actors rely on conflicting theories when making decisions. We accept that these theories are often partially successful in their terms, but also that their application can have unanticipated consequences for their supporters, and undesirable consequences for others.

Our approach is experimental, but we consider our innovations in the methodology of thinking about the future to be demanded by the span and depth of world problems – especially those involving distributional issues. The futures-oriented kind of research presently carried out in international organisations, which are increasingly involved in elaborate global modelling efforts, often fail to confront such issues. While national research projects typically reflect the positions of élite groups in each society, the international agencies' institutionalised research usually plays down international conflicts, excludes domestic ones, and thus provides a lowest common denominator of dominant national positions. While the size of such institutionalised research efforts is often very great (providing them with a sense of authenticity), their scope is very limited.

Our future history may be criticised for being too optimistic in its assumptions. But it can hardly be criticised as a rose-tinted view of the future. Instead, it suggests that a range of likely futures offers very little hope to the poorest groups in the world economy. Even though their absolute circumstances may improve (and we cannot be sure of that), the gaps between rich and poor may well grow far beyond their present appalling levels.

Yet, even with existing technologies, it would be feasible for the world's resources to support all of its present population at relatively high living standards. The amount of effort that would be required to redirect economic activity so as to meet the needs of the poorer half of the world's population in a sustained way, would certainly be immense. But this effort would be nothing compared to the frenetic energy with which nations mobilise in times of warfare, or to the resources dedicated to preparations for war in our uncertain and partial peace. A global, human needs-oriented strategy could abolish the worst extremes of poverty, and substantially reduce global inequalities; but unhappily such a strategy is unlikely to be implemented in the foreseeable future. The obstacles are more social than technical, and necessitate large-scale change in international relations and attitudes.

Our future history does not assume overnight transformations of global strategies. We have attempted to outline future circumstances in which such institutional and economic major changes may take place, but they result from processes which unfold over decades. For the future to offer significantly brighter prospects than those outlined here, significant sections of world opinion must throw themselves behind efforts to effect change sooner rather than later.

To bring about such change is far more than a matter of pulling the perfect strategy out of a hat. It certainly is useful (as in the earlier Bariloche and *World Futures* studies) to demonstrate that, with certain prerequisites, a more equal and high living standard world could be achieved. But no amount of demonstration alone will produce change; and the sorts of action that will be required in order to shift the world towards the necessary conditions in which this can be achieved, are complex. They cannot be summarised in a simple set to guidelines or principles. Social movements in many parts of the world will need to recognise the global dimension of their concerns and struggles in practical terms. The links between the concerns of groups active in peace and disarmament movements, the quality of working life, unemployment, poverty and inequality, women's and minority rights, environment and human dignity, all have to be acknowledged and acted upon. There is no neat strategy for achieving this: all we can do is to reiterate the important role of developing consciousness of the links between, and developing coordination in the actions of, such different movements.

We hope that our study can made some contribution to developing this solidarity; to show that we all have a stake in the future. The dangers of international conflict are high, and in an unequal future they must loom large. The future that we are creating could be hell on earth; the results of major international conflict would affect us all directly, and perhaps fatally. The future is too important for the majority of people to feel that they cannot have a role in shaping it. Thus we have not produced a prediction of what will happen: we have only outlined a future that could happen. We report it here in the hope that it will stimulate people to demand and create a better future for all humanity.

Appendix

THE THEORETICAL STRUCTURE OF THE MODEL

This appendix describes the equations and method of solution for the model used in Chapter 7. This is a modification of the model originally specified by Chichilnisky and described in Chichilnisky and Cole (1978b). Each economy produces three commodities: a basic consumption good, a non-basic or luxury good, and an investment good, all of which are traded internationally. Three factors of production — skilled labour, unskilled labour and capital — are employed in fixed proportions. Supplies of labour are variable and respond to changes in real wage rates. There are two consumer groups: a high-income one which owns the unskilled labour. Ownership of the capital stock is shared.

In equilibrium, demand equals supply for all commodities and factors, there are no excess profits, and the expenditure of each income group is equal to its income. The supply of capital and the levels of investment and net exports are taken to be exogenous, together with parameters describing technology, consumption proportions and labour supplies. The model determines the output of commodities, the level of employment (of labour) and consumption, the balance of trade deficit, and the prices of all commodities and factors in equilibrium.

DEFINITION OF VARIABLES

The endogenous variables Ω_N of the model are:

$\{ x_i \mid i = A, B, I \}$ = gross outputs of commodities

$\{\, p_i \mid i = A, B, I \,\}$ = commodity prices

y_i^l, y^h = consumption levels for low-income and high-income groups, respectively

w^l, w^h = wage rates for unskilled and skilled labour, respectively

π = rate of profit

L, H = supplies of unskilled and skilled labour, respectively

$\{\, b_i \mid i = A, B, I \,\}$ = net exports of commodities

The subscripts ($i = A, B, I$) refer to the luxury, basic and investment goods, respectively. For convenience we suppress the regional index.

The exogenous variables are:

$$\Omega_X = \Big[\, \{\, a_{ij} \mid i, j = A, B, I \,\}, \{\, l_i, h_i, k_i \mid i = A, B, I \,\}, \{\, c_i^l, c_i^h \mid$$

$$i = A, B \,\}, s^l, s^h, \overline{L}, \alpha^l, \overline{H}, \alpha^h, K^l, K^h, i^l, i^h, F^l, F^h \,\Big]$$

where:

$\{\, a_{ij} \mid i, j = A, B, I \,\}$ = intermediate input coefficients

$\{\, l_i \mid i = A, B, I \,\}$ = input coefficients for unskilled labour

$\{\, h_i \mid i = A, B, I \,\}$ = input coefficients for skilled labour

$\{\, k_i \mid i = A, B, I \,\}$ = input coefficients for capital

F = balance of trade deficit

$\{\, c_i^l \mid i = A, B, \,\}$ = consumption coefficients for low-income group

$\{\, c_i^h \mid i = A, B \,\}$ = consumption coefficients for high-income group

s^l, s^h = shares of balance of trade deficit accruing to low-income and high-income groups, respectively

$g^l = - g^h$ = transfers between high and low-income groups

$\bar{L}, \alpha^l$ = supply parameters for unskilled labour

$\bar{H}, \alpha^h$ = supply parameters for skilled labour

K^l, K^h = capital stock owned by low-income and high-income groups, respectively

i^l, i^h = investment undertaken by low-income and high-income groups, respectively

The subscripts ($i = A, B, I$) refer to the three commodities described by the model. For convenience we suppress the regional index.

DEFINITION OF EQUILIBRIUM

Equilibrium is defined as the set of scalars

$$\Omega_N = \left[\{ x_i, p_i, b_i \mid i = A, B, I \}, y^l, y^h, w^l, w^h, \pi, L, H, \right] \quad (1)$$

satisfying the following conditions:

(i) *demand and supply conditions — commodities*

$$X_A = a_{AA}x_A + a_{AB}x_B + a_{AI}x_I + c_A^l y^l + c_A^h y^h + b_A \quad (2)$$

$$X_B = a_{BA}x_A + a_{BB}x_B + a_{BI}x_I + c_B^l y^l + c_B^h y^h + b_B \quad (3)$$

$$X_I = a_{IA}x_A + a_{IB}x_B + a_{II}x_I + i^l + i^h + b_i \quad (4)$$

(ii) *demand and supply conditions — factors*

$$l_A x_A + l_B x_B + l_I x_I = L \quad (5)$$

$$h_A x_A + h_B x_B + h_I x_I = H \quad (6)$$

$$k_A x_A + k_B x_B \ k_I x_I = K^l + K^h \tag{7}$$

(iii) *excess profit conditions*

$$P_A = P_A a_{AA} + P_B a_{BA} + p_I a_{IA} + w^l l_A + w^h h_A + \pi k_A \tag{8}$$

$$P_B = P_A a_{AB} + P_B a_{BB} + p_I a_{IB} + w^l l_B + w^h h_B + \pi k_B \tag{9}$$

$$P_I = P_A a_{AI} + P_B a_{BI} + p_I a_{II} + w^l l_I + w^h h_I + \pi k_I \tag{10}$$

(iv) *budget conditions*

$$\left(p_A c_A^l + p_B c_B^l\right) y + p_I i^l = w^l L + \pi k^l + s^l F + g^l \tag{11}$$

$$\left(p_A c_A^h + p_b c_B^h\right) y^h + p_I i^h = w^h H + \pi K^h + s^h F + g^h \tag{12}$$

(v) *labour supply conditions*

$$L = \bar{L} + \alpha^l w^l / \left(p_A c_A^l + p_B c_B^l\right) \tag{13}$$

$$H = \bar{H} + \alpha^h w^h / \left(p_A c_A^h + p_B c_B^h\right) \tag{14}$$

(vi) *non-negativity conditions*

$$\{x_i, p_i \mid i = A, B, I\}, \ y^l, \ y^h, \ w^l, \ w^h, \ \pi, \ L, \ H, \geq 0 \tag{15}$$

The model is homogeneous in prices and income, so one price variable can be set arbitrarily to determine the scale of the price solution. Conditions (2)–(10) imply that:

$$\left(p_A c_A^l + p_B c_B^l\right) y^l + \left(p_A c_A^h + p_B c_B^h\right) y^h + p_I(i^l + i^h)$$

$$+ p_B b_A + p_B b_B + p_I b_I = w^l L + w^h H + \pi (K^l + k^h) \tag{17}$$

When the balance of trade deficit is distributed to the low-income and high-income groups, i.e.,

$$s^l + s^h = 1 \tag{18}$$

the budget conditions then imply:

$$p_A b_A + p_B b_B + p_I b_I + F = 0 \tag{19}$$

Therefore, each region balances its international payments independently and a solution to the model will satisfy the relevant balance of trade condition. Note that a balance for any subset of five regions implies a balance for the remaining region. Hence one of the equations defining equilibrium can be defined from the others (i.e. the Walras law condition for the model).

SOLUTION ALGORITHM

The equilibrium may be compared using an interative algorithm based on the following sequence of calculations due to Meagher (1980):

(i) assume initial values for commodity prices p_A, p_A, p_B, p_I.

(ii) calculate factor prices:

$$[w^l w^h \; \pi]$$

$$= [p_A \; p_B \; p_I] \begin{bmatrix} (1 - a_{AA}) & -a_{AB} & -a_{AI} \\ -a_{BA} & (1 - a_{BB}) & -a_{BI} \\ -a_{IA} & -a_{IB} & (1 - a_{II}) \end{bmatrix} \begin{bmatrix} l_A & l_B & l_I \\ h_A & h_B & h_I \\ k_A & k_B & k_I \end{bmatrix}^{-1}$$

(iii) calculate labour supplies:

$$L = \overline{L} + \alpha^l w^l / (p_A + c_B^l)$$

$$H = \overline{H} + \alpha^h w^h / (p_A c_A^h + p_B c_B^h)$$

(iv)   calculate gross outputs:

$$
\begin{bmatrix} x_A \\ x_B \\ x_I \end{bmatrix} = \begin{bmatrix} l_A & l_B & l_I \\ h_A & h_B & h_I \\ k_A & k_B & k_I \end{bmatrix}^{-1} \begin{bmatrix} L \\ H \\ K^l + K^h \end{bmatrix}
$$

(v)   calculate consumption levels:

$$
y^l = \left(w^l L + \pi K^l + s^l F + g^l - p_I i^l \right) \Big/ \left(p_A c_A^l + p_B c_B^l \right)
$$

$$
y^h = \left(w^h H + \pi K^h + s^h F + g^h - p_I i^h \right) \Big/ \left(p_A c_A^h + p_B c_B^h \right)
$$

(vi)   calculate excess supplies:

$$
\begin{bmatrix} \Delta A \\ \Delta B \\ \Delta C \end{bmatrix} = \begin{bmatrix} (1 - a_{AA}) & - a_{AB} & - a_{AI} \\ - a_{BA} & (1 - a_{BB}) & - a_{BI} \\ - a_{IA} & - a^{BI} & (1 - a_{II}) \end{bmatrix} \begin{bmatrix} x_A \\ x_B \\ x_C \end{bmatrix} - \begin{bmatrix} c_A^l y^l + c_A^h y^h \\ c_B^l y^l + c_B^h y^h \\ i^l + i^h \end{bmatrix} - \begin{bmatrix} b_A \\ b_B \\ b_I \end{bmatrix}
$$

(vii)   check whether excess supplies are tolerably close to zero; if not revise prices according to an appropriate rule and return to (ii).

In practice, a change in the price of any one commodity significantly affects the excess supplies of all three. Hence, in order to revise the prices in step (vii), it is first necessary to derive a (3 × 3) matrix describing the response of the excess supplies to small changes in the commodity prices. This involves repeated application of a sequence of calculations identical to steps (ii) – (vi). The price changes required to eliminate (or at least reduce, in the case of non-linear responses) the current excess supplies can then be obtained by inverting the (3 × 3) response matrix and post-multiplying by the appropriate vector of excess supply changes. With this arrangement, the algorithm converges satisfactorily.

Bibliography

Acero, L. (1981), 'Workers' Skills as a Critical Issue for Self-Reliance', in Acero *et al.* (1981).

Acero, L., Cole, S. and Rush, H. (1981), *Methods for Development Planning — Scenarios, Models and Micro-Studies*, UNESCO, Paris.

Adelman, I. and Robinson, S. (1978), *Income Distribution Policy in Developing Countries: A Case Study of Korea*, Stanford University Press/World Bank, California.

Auhuwalia, A. *et al.* (1979) 'Growth and Poverty in Developing Countries', *Journal of Development Studies*, vol. 16.

Amin, S. (1979), *Unequal Development*, Harvester Press, Brighton.

Atkinson, A. (1973), *Wealth, Income and Inequality*, Oxford University Press, London.

Averitt, R. (1970), *The Dual Economy*, Norton, New York.

Balassa, B. (1981), *The Newly-Industrialising Countries in the World Economy*, Pergamon Press, Oxford.

Balassa, B. (1982a), 'Disequilibrium Analysis in Developing Countries: An Overview', *World Development*, December.

Balassa, B. (1982b), *Development Strategies in Semi-Industrial Economies*, Johns Hopkins University Press, (for the World Bank) London.

Balassa, B. *et al.* (1971), *The Structure of Protection in Developing Countries*, Johns Hopkins University Press, Baltimore.

Bartsch W. (1977), *Employment and Technology Choice in Asian Agriculture*, Praeger, New York.

Bessant, J. (1980), 'The Influence of Micro-Electronics Technology', in Cole, S. (ed.) (1980b).

Bezdek, R. (1978), 'Postwar Structural and Technological Changes in the American Economy' *OMEGA International Journal of Management Science*, vol. 6, no. 3, pp. 211–25.

Bhagwati, J., Brecher, R. and Hatta, T. (1982), *The Generalised Theory of Transfers and Welfare: Bilateral Transfers in a Multi-Lateral World* (mimeo), Columbia University, New York.

Bigsten, A. (1983), *Income Distribution and Development: Theory, Evidence and Policy*, Heinemann, London.

Bornschier, V. (1983), 'World Economy, Level Development and Income Distribution' *World Development* **II (1)**, 11-20.

Brandt Commission (1980), *North–South: A Programme for Survival*, Pan, London.

Brecher, R. and Bhagwati, J. (1981), 'Foreign Ownership and the Theory of Trade and Welfare' *Journal of Political Economy*, no. 89, pp. 497-511.

Cable, V. (1980) 'Prospects for Economic Cooperation among Developing Countries' *IDS Bulletin* vol. 11 no. 1, pp. 57-63.

Cameron, D. (1978) The Expansion of the Public Sector: a comparative analysis' *American Political Science Review*, vol. 72, no. 4, pp. 1243-61.

Carter, A., (1970), *Structural Change in the American Economy,* Harvard University Press, Cambridge, Mass.

Chapman, J. (1978), 'Are Earnings more Equal Under Socialism? The Soviet Case with some United States Comparisons' in Moroney, J. (1978).

Chichilnisky, G. (1977), 'Economic Development and Efficiency Criteria in the Satisfaction of Basic Needs', *Applied Mathematical Modelling*, vol. 1, no. 6.

Chichilnisky, G. (1980), 'Basic Goods, the Effects of Commodity Transfers and the International Economic Order', *Journal of Development Economics*, pp. 505-19.

Chichilnisky, G. and Cole, S. (1978a) 'Growth of the North and Growth of the South — Some Results on Export-led Growth with Abundant Labor Supply' Harvard Institute for International Development, Discussion Paper no. 42, Boston.

Chichilnisky, G. And Cole S. (1978b) 'A Model of Technology, Distribution and North–South Relations', *Technological Forecasting and Social Change*, vol. 13, no. 4, pp. 297-390.

Chirot, D. (1977), *Social Change in the Twentieth Century,* Harcourt, Brace, Jovanovich, New York.

Clark, A. (1981), 'Third World 2001' *South*, no. 13, November.

Clark, J. and Cole, S. (1978), *Global Simulation Models*, John Wiley, Chichester.

Cline, W. (1982), Can the East-Asian Model of Development be Generalised?, *World Development*, vol. 10, no. 2, pp. 81-90.

Cole, S. (1977), *Global Models and the International Economic Order*, Pergamon Press, Oxford.

Cole S. (ed.) (1980a), *The UNITAR Macro-Model — National Models in the World Economy: Model Structures and Estimation*, no. 1, UNITAR, New York.

Cole, S. (ed.) (1980b), 'Technological Alternatives for Development', *Technical Report*, no. 3, UNITAR, New York.

Cole, S. and Meagher, A. (1981) 'Growth and Income Distribution in India: A General Equilibrium Approach' *Discussion Paper*, 13, August, La Trobe University.

Cole, S. and Miles, I. (1980), 'Labour Supply in the Major World Regions' *UNITAR Working Paper*, Science Policy Research Unit, University of Sussex (mimeo), March.

Cole, S. and Nuñez-Barigga, A. (1981) 'Basic Needs Technology: A Starting-Point for a Macro-Economic Evaluation', *UNITAR Working Paper*, Science Policy Research Unit, University of Sussex (mimeo).

Cole, S. *et al.* (1973), *Thinking about the Future — A Critique of the Limits to Growth*, Sussex University Press, Brighton.

Collard, D., Lecomber, R. and Slater, M. (eds) (1979), *Income Distribution: The Limits to Redistribution*, Halstead Press, Wiley, New York.

Crossman, E. (1966), *Evaluation of Changes in Skill Profile and Job Content due to Technical Change*, University of California, Berkeley.

Cutler, R. M. (1983), 'East–South Relations at UNCTAD: Global Political Economy and the CMEA' *International Organisation*, vol. 37, no. 1, pp. 121-4.

Das Gupta, A. (1976), *A Theory of Wage Policy*, Oxford University Press, London.

De Melo, J. (1978), 'Estimating the Costs of Protection: A General Equilibrium Approach' *Quarterly Journal of Economics*, May, pp. 209-26.

De Melo, J. (1977), 'Distortions in the Factor Market: Some General Equilibrium Estimates' *Review of Economics and Statistics*, November, pp. 398-405.

De Melo, J., Dervis, K. and Robinson, S. (1982) *General Equilibrium Models for Development Policy*, Cambridge University Press, Cambridge.

DHSS (1983), *Family Expenditure Survey*, Department of Health and Social Security, London.

Diamond, Lord (1975, 1976, 1977), *Royal Commission on the Distribution of Income and Wealth*, HMSO, London.

Diwan, D. and Livingstone, D. (1979) *Alternative Development Strategies and Appropriate Technology*, Pergamon Press, Oxford.

Dixit, A. (1983), *The Multi-Country Transfer Problem*, Princeton, N. J. (mimeo), February.

Doblin, C. (1978), *Capital Formation, Capital Stock and Capital Output Ratios (Concepts, Definitions. Data, 1950-1975)*, International Institute for Applied Systems Analysis, Laxenburg, RM–78–70.

Eckaus, R. S. *et al.* (1976) *Multi-Sector Equilibrium Policy Models for Egypt* (mimeo), MIT, Cambridge, Mass.

Evans, D. (1984), in Kaplinsky, R.

Evans, E. and Cole, S. (1979), 'The Empirical Formulation of the Multi-Region UNITAR Model' in Cole, S. (ed.) (1980).

Felix, D. (1977), 'The Technological Factor in Socio-Economic Dualism: Toward an Economy of Scale Paradiom for Development Theory' *Economic Development and Cultural Change* vol. 25 (supplement), pp. 180-211.

Fields, G. (1980), *Poverty, Inequality and Development*, Cambridge University Press, London.

Foxley, A. (1976), 'Redistribution of Consumption: Effects on Production and Employment' *Journal of Development Studies*, **12 (3)**, April.

Freeman, C. and Jahoda, M. (1978), *World Futures: The Great Debate*, Martin Robertson/Universe, London and New York.

Frey, B. (1978), *Modern Political Economy*, Martin Roberston, Oxford.

Fry, M. (1981), *Interest Rates in Asia*, University of Hawaii (mimeo), referenced in Balassa, B. (1981).

Gaiha, R. (1979), 'On Testing the Stability of Input–Output Relations in the Indian Economy' *Journal of Development Economics*, no. 7.

Gale, D. (1974), 'Exchange Equilibrium and Coalitions' *Journal of Mathematical Economics*, vol. 1, pp. 63-6.

GATT (1978), *International Trade 1977/78*, Geneva.

Gershuny, J. I. (1979), 'The Informal Economy', *Futures*, February.

Godet, M. (1978), *The Crisis in Forecasting and the Emergence of the 'Prospective' Approach'*, Pergamon Press, New York.

Griffin, R. (1981), *Land Concentration and Rural Poverty*, Macmillan, London.

Guetzkow, H. and Valedez, J. (eds) (1981), *Simulated International Processes*, Sage, Beverly Hills, Cal.

Gupta, A. (1975), 'The Rich, the Poor and the Taxes they pay in India' *World Employment Programme*, WEP 2–12, International Labour Organisation, Geneva.

Guttman, P. (1977), 'The Subterranean Economy', *Financial Analysts' Journal*, December.

Hallak, J. (1978), *Education and Work in Indonesia*, International Institute for Education and Development, Paris.

Hallak, J. and Caillods, F. (1979), 'Education, Training and the Traditional Sector', IIEP, *Fundamentals*.

Hayden, C. and Round, J. (1983), 'Analytic Foundations of Planning Exercises based on a SAM Framework', Development Economics Research Centre Discussion Paper, Warwick.

Hazlehurst, R. *et al.* (1969), 'A Comparison of the Skills of Machinists on Numerically-Controlled and Conventional Machines', *Occupational Psychology*, vol. 43, nos 3 & 4.

Helleiner, G. (1973), 'Manufactured Exports from Less-Developed Countries and Multinational Firms', *Economic Journal 83* (March).

Herrera, A. *et al.* (1976), *Catastrophe or New Society?*, IDRC, Ottawa.

Hillman, M. and Whalley, A. (1980), *The Social Consequences of Rail Closures*, Policy Studies Institute, London.

Hinchcliffe, K. (1975), 'Education, Individual Earnings and Earnings Distribution' *Journal of Development Studies*, vol. 11, no. 2, January.

Hird, C. and Irvine, J. (1979), 'The Poverty of Wealth Statistics' in *Demystifying Social Statistics*, Irvine, J., Miles, I., and Evans, J., (eds), Pluto, London.

HMSO (1980) *Social Trends*, No. 9, Central Statistical Office, HMSO, London.

Hopkins, M. and Van der Hoeven, R. (1983), *Basic Needs in Development Planning*, Gower Press, Aldershot.

Horowitz, M. *et al.* (1966), *Manpower Requirements for Planning: An International Comparison*, Northeastern University, Boston.

Hufbauer, G. C. (1970), 'The Impact of National Characteristics and Technology in the Commodity Composition of Trade in Manufactured Goods', in Vernon, R. (ed.) (1970).

IBRD (1978:80), *World Development Reports 1977 to 1979*, and *Annex on World Development Indicators*, IBRD, Washington.

IMF (1983), *World Economic Outlook 1982-83*, International Monetary Fund, Washington, April.

Independent Commission on International Development Issues (1979), *North–South: A Programme for Survival*, Pan London. (Brandt Commission)

International Labor Organisation (1978), *ILO Yearbook*. ILO, Geneva.

Jahoda, M. (1973), 'Postscript', in Cole, S. *et al.* (eds) (1973).

Johnson, H. (1966), 'Factor Market Distortions and the Shape of the Transformation Curve', *Econometrics*, July, pp. 686-968.

Jones, R. (1982a) *The Transfer Problem in the Three-Agent Setting* (mimeo), Rochester University, NY.

Jones, R. (1982b), *Income Effects and Paradoxes in the Theory of International Trade* (mimeo), Rochester University, NY.

Kahn, H. and Wiener, A. (1967), *The year 2000*, Macmillan, London.

Kaminsky, B. and Okolski, M. (1981), *An Estimation of the UNITAR Model for Poland*, Faculty of Economics, University of Warsaw (mimeo), July.

Kaplinsky, R. (1984), (ed.) 'International Context for Industrialisation in Coming Decades', *Journal of Development Studies*, vol. 24, in press.

Kaplinsky, R. (1981), 'Inappropriate Products and Techniques in UDCs: The Case of Breakfast Foods in Kenya', in Acero, *et al.* (1978).

Kaplinsky, R. (1983), *Automation in a Crisis*, Longmans, London.

Kay, J. (1979) 'The Anatomy of Tax Avoidance', in Collard, D. *et al.* (1979).

Keesing, D. (1966) 'International Economic: Progress and Transfer of Technical Knowledge, Labor Skills and Comparative Advantage', *American Economic Review*, May.

Kidron, M. and Segal, R. (1981) *The State of the World Atlas*, Pluto Press, London.

Kittler, F. (1979) *International Trade Matrix 1962-77: Estimation of Trade Flows in the Unitar Six Region Model* (mimeo), Columbia University/UNITAR.

Kravis, I., Heston, A., and Sommers, R. (1982), *World Product and Income: International Comparisons of Real Product*, Johns Hopkins University Press, Baltimore.

Kuznets, S. (1955), 'Economic Growth and Income Inequality', *American Economic Review*, vol. 45.

Kuznets, S. (1963), 'Quantative Aspects of the Growth of Nations', *Economic Development and Cultural Change*, vol. II.

Lall, S. (1980) 'Developing Countries as Exporters of Industrial Technology', *Research Policy*, no. 9, pp. 24-52.

Lane, D. (1971), *The End of Inequality*, Penguin, Harmondsworth.
Layard, P. and Saigal, J. (1966), 'Educational and Occupational Characteristics of Manpower: An International Comparison', *British Journal of Industrial Relations*, vol. IV, July.
Lederer, K. (ed) (1980), *Human Needs: a contribution to the current debate* Oelgeschlage, Cuinn and Main, Cambridge, Mass.
Leontieff, W. *et al.* (1976), *The Future of the World Economy*, United Nations, New York.
Lewis, W. A. (1954), *Economic Development with Unlimited Supplies of Labour*, The Manchester School, May.
Lluch, C., Powell, A. and Williams, R. (1977), *Patterns in Household Demand and Saving*, Oxford University Press/World Bank, Washington.
Lysy, F. and Taylor, L. (1979), 'Vanishing Short-Run Income Redistributions: Keynesian Clues about Model Surprises', *Journal of Policy Modelling*.
Maslow, A. (1954), *Motivation and Personality*, Harper, New York.
Meadows, D. *et al.* (1972), *The Limits to Growth*, Universe Books, New York.
Meagher, A. (1980), 'Empirical Estimation and Aggregation for a computable Global General Equilibrium Model', in Cole, S. (ed.) (1980).
Meagher, A. and Cole, S. (1981), 'The Empirical Estimation of a Global Economic Model' in Acero, *et al.* (eds) (1981).
Mesarovic, M. and Pestel, E. (1974), *Mankind at the Turning Point*, Dutton/Readers Digest, New York.
Miles, I. (1983), 'New Technologies, Old Orders' in Masini, E. (ed.), *Visions of Desirable Societies* Oxford: Pergamon.
Miller, S. (1975) 'Notes on Neo-Capitalism' *Theory and Society* vol. 2, no. 1, 1-35.
Murray, R. 1971, 'The Internationalisation of Capital and the Nation State', *New Left Review* no. 67, pp. 84-109.
Moroney, J. (1978), *Income Inequality: Trends and Comparisons*, Lexington Books, Toronto.
Myrdal, G. (1970), *The Challenge of World Poverty*, Parthenon Books, New York.
Nayyar, D. (1978a), 'Transnational Corporations and Manufactured Exports from Poor Countries', *Economic Journal*, vol. 88 (March).
Nayyar, D. (1978b), 'Industrial Development in India: Some Reflections on Growth and Stagnation', *Economic and Political Weekly*, Special under vol. 13, nos. 31-3, pp. 1265-78.

Nuti, D. (1979) 'The contradictions of Socialist Economies' in Miliband, R. and Saville, J. (eds), *The Socialist Register 1979*, Merlin, London.

OECD (1970), *Gaps in Technology: Analytical Report*, OECD, Paris.

OECD (1979), 'Capital Goods: Structural Evolution and World Prospects', FUT (79) C.4, OECD, Paris.

OECD Interfutures (1979), *Facing the Future: Mastering the Probable and Managing the Unpredicatable*, OECD, Paris.

OECD (1980), *Economic Outlook*, OECD, Paris, December.

Onishi, A. (1983) 'The Fugi Macro-Economic Model and World Trade to 1990', *Futures*, April, pp. 99-110.

Paukert, F. (1973), 'Income Distribution at Different Levels of Development: A Survey of Evidence', *International Labor Review*, **108 (2-3)**, pp. 97-121.

Pavitt, K. (1980), 'Technical Innovation and Industrial Development: The Dangers of Divergence', *Futures*, February.

Petras, J. (1975) 'Socialist Revolutions and their Class Components' *New Left Review* no. 111, pp. 37-64.

President's Commission on the Year 2000 (1980), *The Global 2000 Report to the President's Department of State*, Washington, DC.

Pyatt, G. and Roe, A. (1979) *Social Accounting for Development Planning with Special Reference to Sri Lanka*, IBRD, Washington.

Pyatt, G. and Round, J. (1979) 'Social Accounting Matrices for Development Planning', *The Review of Income and Wealth*, Series 23, no. 4, December.

Rada, J. (1980), *The Impact of Micro-Electronics*, ILO, Geneva.

Ranis, G. (1983) 'The NICs, the near-NICs, and the World Economy' (mimeo), presented at seminar at Institute of Development studies , University of Sussex.

Rodgers, G., Hopkins, M. and Wery, R. (1979), *Population, Employment and Inequality: Bachue-Philippines*, Saxon House, Westmead.

Round, J. and Hayden, C. (1980), *Development in Social Accounting Methods as Applied to the Analysis of Income Distribution and Employment Issues*, Zimbabwe Economics Symposium, 1980.

Round, J. (1982), 'Economy Wide Multipliers and Project Appraisal' (mimeo) Department of Economics, University of Warwick.

Ruffing, L. T. (1980), *Review of Data Available for Analysing Occupational Structure and Educational Attainment in Fourteen Developing Countries*, International Labour Organisation, Geneva.

Rutherford, M. (1978) 'Economic Growth by Moonlight', *Business Week*, March.

Samuelson, P. (1971), 'On the Trail of Conventional Beliefs About the Transfer Problem', in Bhagwati, J. *et al.* (eds), *Trade, Balance of Payments and Growth*, Amsterdam, North-Holland.

Sanders, C. (1980) 'Measures of Total Household Consumption, *The Review of Income and Wealth*, no. 4.

Sarabhai, V. (1968), 'United Nations Conference on Space Research', cited in Clark, A. (1981).

Scammel, W. (1980), *The International Economy since 1945*, Macmillan, London.

Segal, R. and Kidron, M. (1980), *The State of the World Atlas*, Pan/Pluto, London.

Singer, H. (1977), *Technologies for Basic Needs*, International Labor Organisation, Geneva.

Singer, H. and Ansari, J. (1978), *Rich and Poor Countries*, George Allen & Unwin, London.

Singh, A. (1979), 'A Basic Needs Approach to Development versus the New International Economic Order', *World Development*, vol. 7 pp. 585-606.

Snyder, D. and Kick, K. (1979), 'Structural Position in the World Economy and Economic Growth, 1955-1970' *American Journal of Sociology*, vol. 84, no. 5, pp. 1096-126.

Soltow, C. (1968), 'Long-Run Changes in British Income Inequality', *Economic History Review*, 21 (1), 17-29, reprinted at Atkinson, A. (1973), *Wealth, Income and Inequality*, Oxford University Press, London.

Squire, L. (1979), *Labor Force Employment and Labor Markets in the Course of Economic Development*, World Bank Staff Working Paper, no. 336, New York.

Steinberg, E. and Reger, J. (1978), *New Means of Financing International Development Needs*, Brookings Institution, Washington.

Stephens, J. (1979) *The Transition from Capitalism to Socialism* Macmillan, London.

Stewart, F. (1978), *Technology and Underdevelopment*, 2nd edn, Macmillan London.

Stewart, F. and James, J. (eds) (1982), *The Economics of New Technology in Developing Countries*, Francis Pinter/Westview, London.

Summers, L. (1981), 'Capital Taxation and Accumulation in a Life Cycle Growth Model', *American Economic Review*, September pp. 533-44.

Sutcliffe, R. (1971), *Industry and Underdevelopment*, Addison-Wesley, New York.

Syriquin, M. (1973), 'Efficient Input Frontiers for the Manufacturing Sector in Mexico 1965-1980', *International Economic Review, October*.

Szakolczai, G. (1979), 'Limits to Redistribution: The Hungarian Experience', in Collard, D. *et al.* (eds) (1979).

Taylor, L. (1980), *Macro-Models for Developing Countries*, McGraw-Hill, New York.

Thirsk, W. (1980), 'Aggregation Bias and the Sensitivity of Income Distribution to Changes in the Composition of Demand: The Case of Columbia', *Journal of Development Studies*. vol. 18.

Thurow, L. (1983), 'A Rising Tide of Poverty', *Newsweek*, 11 July, p. 62.

UNCTAD (1979), 'Trade and Development', *An UNCTAD Review*, no. 1, Geneva.

UNCTAD (1981), *Trade and Development Report*, United Nations, New York.

UNESCO (1981), cited in Acero, L. *et al.* (1981).

UNIDO (1978), *Capital Goods Industry — Preliminary Study*, United Nations Industrial Development Organisation, Vienna.

UNITAD (1981), *Reports on the UNITAD Model: ACC Working Group to Committee for Development Planning* (mimeo), United Nations, Geneva.

UNSO (1968) *A System of National Accounts*, Series F, no. 2, rev. 3, United Nations, New York.

United Nations (1975–7), *Yearbook of National Account Statistics*, vols I & II, *Individual County Data*; vol. III, *International Accounts Statistics*, New York.

United Nations (1975–80), *Statistical Yearbook*, United Nations, New York.

United Nations (1979), *Employment, Income Distribution and Consumption: Long-Term Objectives and Structural Changes*, United Nations, New York.

United Nations (1982), *World Economic Survey 1981-82*, United Nations, New York.

United Nations (1983), *Overcoming Economic Disorder: United Nations Committee for Development Planning*, United Nations, New York.

United States Department of Labor (1970), *Directory of Occupational Titles*, Columbia University Press, New York.

Vernon, R. (ed.) (1970), *The Technology Factor in International Trade*, Columbia University Press, New York.

Vitelli, G. (1980), *The Chaotic Economics of Technical Change: A Survey Throughout the Choice of Technique*, Institute for Development Studies, University of Sussex (mimeo).

Ward, B. *et al.* (1970), *The Widening Gap: Development in the 1970s*, Columbia University Press, New York.

Ward, M. (1981), *Limitations of the Social Accounting Matrix Approach for Development Policy and Distributional Analysis*, 7th General IARIW Conference 1981, Institute for Development Studies, Sussex (mimeo).

Weintraub, S. (1979) 'The New International Order: the beneficiaries' *Word Development*, vol. 7, no.3, pp. 247-58.

World Bank (1983), *Development Reports 1980-1983*, IBRD, Washington.

World Bank/IBRD (1977–83), *World Development Reports and Annex on World Development Indicators*, IBRD, Washington.

Index